STATISTICS IN THE
SOCIAL SCIENCES

STATISTICS IN THE SOCIAL SCIENCES
Current Methodological Developments

Edited by

Stanislav Kolenikov
Douglas Steinley
Lori Thombs
University of Missouri–Columbia

WILEY

A JOHN WILEY & SONS, INC., PUBLICATION

Published by John Wiley & Sons, Inc., Hoboken, New Jersey.
Published simultaneously in Canada.

For general information on our other products and services or for technical support, please contact our Customer Care Department within the United States at (800) 762-2974, outside the United States at (317) 572-3993 or fax (317) 572-4002.

Wiley also publishes its books in a variety of electronic formats. Some content that appears in print may not be available in electronic format. For information about Wiley products, visit our web site at www.wiley.com.

Library of Congress Cataloging-in-Publication Data:

Statistics in the social sciences: current methodological developments /
    edited by Stanislav Kolenikov, Douglas Steinley, Lori Thombs
    Includes bibliographical references and index.
    ISBN 978-0-470-14874-7

10 9 8 7 6 5 4 3 2 1

CONTENTS

LIST OF FIGURES

LIST OF TABLES

PREFACE

This book is aimed at a wide spectrum of researchers and students who are concerned with current statistical methodology applied in the social sciences. In short, all social scientists working in fields including (but not limited to) psychology, sociology, test theory, market research, and many more would benefit from being kept abreast of the cutting-edge research contained herein.

The impetus for the book was *The Sixth Annual Winemiller Conference: Methodological Developments of Statistics in the Social Sciences* held at the University of Missouri in Columbia, Missouri from October 11 to 14 in 2006. The aim of the conference was to foster collaboration among mathematical statisticians and quantitatively oriented social science researchers. This interdisciplinary conference brought top researchers from major social sciences disciplines, highlighting the interface between recent developments in each area. The idea was to gather experts in the field and to assemble and edit a text that encompassed many of the current methodologies applied by researchers in different social sciences disciplines.

Chapter 1's devoted to structural equation modeling is by Peter Bentler and Victoria Savalei. Their focus is, however, on correlation structures. They describe the typical problems that arise in analysis of correlations and correlation matrices. From the field of structural equation and covariance structure modeling, they borrow the estimation methods and test statistics commonly used in SEM, as well as corrections commonly used to improve the finite sample behavior of those methods, and draw parallels in correlation structure analysis. Bentler and Savelei begin by providing different parameterizations of correlation matrices from sturctural equation models in Jöreskog's LISREL formulation and in the Bentler–Weeks formulation. Then they highlight the differences between covariance (or mean-and-covariance) structure and correlation structure analyses, show where each is best applied, and demonstrate why treating correlation matrices as covariance matrices may lead to erroneous inference. Asymptotic theory under correct specificaiton is derived for estimation of structural correlation models based on the general concept of discrepancy functions. Quadratic forms in parameter estimates are shown to be the generic asymptotic form of such discrepancy functions, and further analysis proceeds in terms of those forms leading to (asymptotically distribution free, ADF) M-estimates of the structural parameters. Asymptotic normality of those estimates is demonstrated through the standard arguments, and considerations of asymptotic efficiency via the optimal choice of the weighting matrix are given. To obtain asymptotic efficiency, the weight matrix in the quadratic form minimization problem should be chosen to be the asymptotic variance of the sample correlations, the general form of which is given. Misspecification of the model structure and weight matrix is then considered, leading to more complicated distributions of the goodness-of-fit statistics (noncentral χ^2 and mixtures of χ_1^2's). Other limitations of the ADF estimation methods, such as indomitable sample-size requirements, are discussed and some remedies proposed that might improve the finite sample performance of the goodness-of-fit tests, similar to the way that those corrections operate in classical covariance structure SEMs.

When the weight matrix is misspecified, or computation of the optimal matrix is not feasible (e.g., in large models with dozens of variables), the distribution of the goodness-of-fit test statistics becomes a mixture of χ_1^2's with potentially unequal weights. Bentler and Savalei provide several ways of approaching that distribution. The weights can be estimated from the existing parameter estimates, leading to a complicated problem of finding the quantiles of the resulting distribution. Satterthwaite-type corrections (known as Satorra–Bentler scaled and adjusted statistics in SEM) can be applied. Or, alternatively, an entirely different statistic with asymptotic χ^2 distribution can be constructed from correlation residuals rather that from the discrepancy functions. Some additional simplification of the analysis is feasible when specific distributions of the data can be assumed. Under normality, more stucture can be found in the variance of the sample correlations, and hence the weight matrix of the ADF, leading to an analytically feasible inverse of the weight matrix. The normality assumption can be relaxed somewhat to the assumption of elliptically contoured distributions, and the only modification that needs to be made to the normal theory methods is scaling by the common kurtosis. A further step down from distributional assumptions might be heterogeneous kurtosis theory, which has not yet received much attention in the literature. Bentler and Savalei exemplify their ideas with a classic anthropometric example with eight physical measurement variables and a small simulation. A two-factor model with five variables per factor was used in simulations, and Vale–Maurelly transformation was used to make data nonnormal. It was found that while the ADF method that was applied to either covariance or correlation matrix was overrejecting for most sample sizes, the ADF with structured correlation matrix mildly underrejected at moderate sample sizes, attaining its asymptotic test size with samples of size 400 and above. It was also found that the residual-based test statistic and residual-based F-statistic behaved rather poorly in unstructured ADF correlation analysis, while the Satorra–Bentler scaled statistic yielded rejection rates closer to the nominal. All statistics improved considerably, however, when a structured weight matrix was used in ADF. Thus, Bentler and Savalei have proposed a number of approaches to the analysis of correlation structures. Some of those proposals are completely new, and most of the others have received very limited attention in the social science statistics literature. It might then be expected that this chapter will provide a fertile field for research that can provide further analytical, simulation and empirical support to Bentler and Savalei's ideas.

Chapter 2 is by Kenneth Bollen, Daneil Bauer, Sharon Christ, and Michael Edwards, who review the area of structural equation modeling. They give a brief review of the history of the field, introduce the basic ideas and notation, and demonstrate how general SEMs specialize to such special cases as simultaneous equation models in econometrics, multiple regression and ANOVA, and confirmatory factor analysis. In general, structural equation modeling would proceed by specifying the model, computing the implied moment matrices, establishing identification, estimating parameters and assessing the fit of the model, with additional respecification if the model fits poorly. Bollen et al. consider those steps one by one, briefly discussing the procedures commonly used. They present the maximum likelihood estimator, the two-stage instrumental variable estimator, and the least-squares/asymptotically distribution-free estimator commonly used by applied researchers. Having outlined the general modeling steps, Bollen et al. proceed to discuss several recent extensions. One of them is the hybrid structural equation and multilevel modeling. A common way to analyze multilevel SEM is to specify a parametric model for both within- and between-group covariance matrices, which makes it possible to model the contextual and individual effects, just as in traditional linear multilevel models. Moreover, not only the parameter values, but even the factor structure, can be specified differently for the within- and between-group parts of the model. Another view of multilevel SEM is to explicitly specify the higher-level random effects as latent factors. Similar ideas have long been used in growth curve modeling, and the synthesis with SEM has been proposed recently, as reviewed in Section 2.2.

Another hybrid type of modeling arises when the structural equation models are crossed with latent class models, giving rise to structural equation mixture models. Those models are also related to the item response theory models operating on discrete outcomes, to growth mixture models, and to nonparametric maximum likelihood estimators of SEM that do not specify distribution of the latent variables but, rather, estimate it. Bollen et al. further discuss the issues of identification and sensitivity to assumptions, which become more acute in those more complicated models. A number of interesting applications are considered, from direct class modeling to semiparametric nonlinearity modeling and semiparametric modeling of the latent variable distributions. In the next section of the chapter they review the relation of SEM to item response models, some forms of which can be cast as confirmatory factor analysis with categorical variables. Complications arising from the discrete nature of the data are discussed, and estimation methods reviewed. The last extension of the SEM discussed by Bollen et al. is to complex samples. An overview of the basic complex sample design features, such as clustering, stratification, and unequal probabilities of selection, is given. It is shown how those features violate the model-based SEM assumption, and how estimation procedures are then affected. Sample weights are motivated through a Horvitz–Thompson estimator of a total. An applied researcher can then proceed by attempting to model the sample design with, say, random effects for clusters and categorical variables for strata; or one can use estimation procedures that correct for the complex survey design, such as weighted estimation and pseudo-maximum likelihood. Special care should be taken to estimate the variances of the parameter estimates properly, through either a sandwich-type estimator or through appropriate survey design resampling schemes.

In **Chapter 3**, Lawrence Hubert, Hans-Friedrich Köhn, and Douglas Steinley discuss strategies for the hierarchical clustering of an object set to produce a sequence of nested partitions in which object classes within each successive partition are constructed from the union of classes present at the previous level. In turn, any such sequence of nested partitions can be characterized by what is referred to as an ultrametric, and conversely, any ultrametric generates a nested collection of partitions. There are three major areas of concern in this paper: (1) the imposition of a given fixed order, or the initial identification of such a constraining order, in constructing and displaying an ultrametric; (2) extensions of the notion of an ultrametric to use alternative collections of partitions that are not necessarily nested but which do contain objects within classes consecutive with respect to some particular object ordering. A method for fitting such structures to a given proximity matrix is discussed along with an alternative strategy for graphical representation; (3) for the enhanced visualization of additive trees, the development of a rational method of selecting a root by imposing some type of order-constrained representation on the ultrametric component in a decomposition of an additive tree (nonuniquely into an ultrametric and a centroid metric). A simple numerical example will be used throughout the paper based on a data set characterizing the agreement among the Supreme Court Justices for the decade of the Rehnquist Court. All the various MATLAB M-files used to illustrate the extensions are available as open-source code from a web site given in the text.

In **Chapter 4**, Michael Brusco, Stephanie Stahl, and Dennis Cradit discuss using multidimensional scaling (MDS) in the city-block metric, an important tool for representing the psychological space associated with separable stimuli. When two or more proximity matrices are available for the same set of stimuli, the development of a city-block MDS structure that fits each of the matrices reasonably well presents a challenging problem that might not be solved by pooling the data (e.g., averaging) across matrices. These authors present a multiobjective programming approach for multidimensional city-block scaling of multiple proximity matrices. The multiobjective function of the model is composed of either weighted least-squares loss functions or, in cases where nonmetric relaxations of the proximities are desired, weighted stress functions. The multiobjective function is optimized subject to constraints on the permutation of the objects on each dimension. Because there are well-noted problems with gradient-based approaches for city-block MDS, a combinatorial heuristic procedure is proposed for solving the multiobjective programming city-block model. The model is demonstrated using empirical data from the psychological literature.

In **Chapter 5**, Jeff Gill compares Bayesian and frequentist approaches to estimation and testing of social science theories. He first argues why fixing the data and conditioning on them, as is done in Bayesian statistics, is a reasonable starting point in social sciences: indeed, the repeated sampling necessary to justify the frequentist paradigm is hardly feasible with constantly changing social and human environments. Upon providing the mechanics of Bayes theorem and Bayesian inference, he considers a small example with count data, and demonstrates graphically the process of prior updating. The choice of the prior is further provided. Differences in how the models are set up, and how analysis then proceeds and inference is conducted, are highlighted between the Bayesian and the frequentist paradigms, with somewhat provocative comparisons between the two paradigms and dominant data analysis standards. Then Gill reviews the existing approaches to hypothesis testing and shows step-by-step procedures in the Fisher paradigm, Neyman–Pearson paradigm, Bayesian paradigm, and the null hypothesis significance testing paradigm. Gill's argument against the latter is supported by several dozen references in statistics and social and behavioral sciences. He then comes back to the counts example and shows an extension of his analysis to a (rather difficult, in any paradigm) problem of change-point estimation. He shows how a Gibbs sampler can be set up for this problem by explicitly specifying the full conditional distributions, and how convergence of the resulting Markov chain can be established. He then reviews the substantial results and notes that the Bayesian estimates of the change point are the probable cause of the change. He concludes by highlighting again the critical differences between Bayesian and frequentist paradigms, and provides philosophical considerations for the former.

In **Chapter 6**, Jeff Rouder, Paul Speckman, Douglas Steinley, Michael Pratte, and Richard Morey show how the shape of a response-time distribution provides valuable clues about the underlying mental processing. If a manipulation affects the shape of an RT distribution, it is reasonable to suspect that the manipulation has done more than simply speed or slow the rate of processing. They develop a nonparametric bootstrap test of shape invariance. Simulations reveal that the test is sufficiently powered to detect small shape changes in reasonably sized experiments while maintaining appropriate type I error control. The test is simple and can be applied broadly in cognitive psychology. An application to a number priming experiment provides a demonstration of how shape changes may be detected.

In **Chapter 7**, Joseph Hilbe outlines the standard computer programs used for statistical analysis and emphasizes those that should get more use. The packages include R, SAS, SPSS, STATISTICA, Stata, StatXact/LogXAct, Stat/Transfer, ePrint Professional, and nQueary Advisor.

The 2006 Winemiller Conference and this book would not have been possible without the generous support of Albert Winemiller, whom we would like to thank for his interest in cultivating the integration of mathematical statistics and social science at both the theoretical and applied levels. We would also like to thank the Department of Statistics (and its Chair, Dr. Nancy Flournoy) for sponsoring the conference and providing support for Dr. Thombs and Dr. Kolenikov in the editing of this book. We also thank the Department of Psychological Sciences (and its Chair, Dr. Ann Bettencourt) for providing support for Dr. Steinley throughout the editorial process.

We express our deep gratitude to all of the authors for their contributions and their patience in the process of bringing this book together. We also extend our thanks to the other two members of the organizing committee: Dr. Nancy Flournoy and Dr. Steve Osterlind. Additionally, for their work on the 2006 Winemiller Conference, we acknowledge the assistance of Ray Bacon, Margie Gurwitt, Gretchen Hendrickson, Wang Ze, and Peggy Bryan. This conference was supported by the National Science Foundation under Grant 0605679. Finally, we thank John Wiley & Sons, executive editor Steve Quigley, and project manager Jacqueline Palmieri for supporting publication of the book.

Stanislav Kolenikov
Douglas Steinley
Lori Thombs

ABOUT THE AUTHORS

Daniel J. Bauer is an Associate Professor in the Quantitative Psychology Program of the L. L. Thurstone Psychometric Laboratory in the Department of Psychology at the University of North Carolina, Chapel Hill. Additionally, Dr. Bauer is currently on the editorial boards of *Psychological Assessment* and *Psychological Methods*. His research focuses on the development, evaluation, and application of quantitative methods (including multilevel linear/nonlinear models, structural equation models, growth mixture models, etc.) suited to the study of developmental phenomena, especially social development in the domains of aggression, antisocial behavior, and substance use.

Peter M. Bentler is a Distinguished Professor of Psychology and Statistics at the University of California, Los Angeles and Director of the UCLA Center for Collaborative Research on Drug Abuse. Additionally, Dr. Bentler serves on the editorial boards of *Applied Psychological Measurement, Archives of Sexual Behavior, Computational Statistics & Data Analysis, Journal of Health Psychology, Multivariate Behavioral Research, Prevention Science, Psychometrika, Sociological Methods & Research*, and *Structural Equation Modeling*. His research focuses on multivariate statistics with emphasis on latent variable and structural equation modeling.

Kenneth A. Bollen is Director of the H. W. Odum Institute for Research in Social Science and the Henry Rudolph Immerwahr Distinguished Professor in the Department of Sociology at the University of North Carolina, Chapel Hill. Furthermore, Dr. Bollen is also currently on the editorial boards of *Multivariate Behavioral Research, Social Forces, Sociological Methods & Research, Structural Equation Modeling*, and *Studies in Comparative International Development*. While his interests in statistical methods are focused primarily on structural equation models, Professor Bollen is also interested in international development, comparative political structures, and natinonal democratic political structures.

Michael Brusco is the Synovus Professor of Business Adminstration at Florida State University. Dr. Brusco's research focuses on cluster analysis, seriation, regression, permutation tests, scheduling, and facility layout. Additionally, he is on the editorial boards of *Journal of Classification* and *Journal of Problem Solving*.

Sharon L. Christ is a Statistical Analyst at the H. W. Odom Institute for Research in Social Science.

J. Dennis Cradit is Dean of the College of Business at Southern Illinois University. Dr. Cradit's research interests focus on marketing segmentation, quantitative methods, and business-to-business marketing.

Michael C. Edwards is an Assistant Professor in Quantitative Psychology at The Ohio State University. His primary research focuses on item response theory and factor analysis, with specific interest in using Markov chain Monte Carlo estimation for multidimensional item response theory.

Jeff Gill is a Professor in the Department of Political Science and Director of the Center for Applied Statistics at Washington University. Professor Gill is on the editorial boards of *Journal of Public Administration Research and Theory*, *Political Analysis*, and *State Politics and Policy Quarterly*. His research applies Bayesian modeling and data analysis (decision theory, testing, model selection, elicited priors) to questions in general social science quantitative methodology, and American political behavior and institutions, focusing on Congress, the bureaucracy, and voters, using computationally intensive tools (Monte Carlo methods, MCMC, stochastic optimization, nonparametrics).

Joseph Hilbe is an Adjunct Professor of Statistics at Arizona State University and is a Fellow of both the American Statistical Association and Royal Statistical Society. Professor Hilbe serves as the Software Reviews Editor for *The American Statistician*.

Lawrence Hubert is the Lyle H. Lanier Professor of Psychology, Professor of Statistics, and Professor of Educational Psychology at the University of Illinois, Champaign-Urbana. Professor Hubert is currently on the editorial boards of *British Journal of Mathematical and Statistical Psychology* and *Journal of Classification*. His research includes data analytic methods focused on data representation and strategies of combinatorial data analysis, including exploratory optimization approaches and confirmatory nonparametric methods.

Hans-Friedrich Köhn is an Assistant Professor of Quantitative Psychology at the University of Missouri. His research interests include multidimensional scaling, combinatorial data analysis, and structural representations of proximity data.

Stanislav Kolenikov is an Assistant Professor of Statistics and an Affiliate Faculty at the Social Sciences Statistics Center of the University of Missouri. His research interests include structural equation modeling, complex survey data analysis, and spatial statistics.

Richard D. Morey is an Assistant Professor of Psychometrics and Statistical Techniques at University of Groningen, The Netherlands.

Michael S. Pratte is a graduate student in cognitive psychology at the University of Missouri.

Jeffrey N. Rouder is a Professor of Cognitive Psychology at the University of Missouri. Professor Rouder research concerns developing nonlinear hierarchical models for improved estimation and inference and studying memory, attention, perception, categorization, and letter recognition.

Victoria Savalei is an Assistant Professor of Quantitative Psychology at the University of British Columbia. Her research interests are structural equation and latent variable modeling.

Paul L. Speckman is Professor of Statistics at the University of Missouri. Professor Speckman is a Fellow of the American Statistical Association and the Institute of Mathematical Statistics. His research concerns nonparametric curve estimation, Bayesian nonparametrics, Bayesian computation, and the application of statistics in cognitive research in psychology.

Douglas Steinley is an Associate Professor of Quantitative Psychology at the University of Missouri and an Affiliate Faculty at the Social Science Statistics Center at the University of Missouri. He is currently the Book Review Editor for the *Journal of Classification*, and his research focuses on multivariate analysis; specific interests include cluster analysis, data reduction, and social network analysis.

Lori A. Thombs is an Associate Professor of Statistics and Director of the Social Sciences Statistics Center at the University of Missouri, Columbia. Her research interests include linear and nonlinear time-series analysis, predictive modeling and classification techniques, interdisciplinary statistical methods, and innovation in statistical education. She is past Associate Editor of *The*

American Statistician and *Journal of Statistical Computation and Simulation*, and has been a member of several NSF review panels and referee for many major statistical journals.

CHAPTER 1

ANALYSIS OF CORRELATION STRUCTURES: CURRENT STATUS AND OPEN PROBLEMS

1.1 INTRODUCTION

The analysis of correlations among variables played an important part in the growth of psychometrics and statistics in the early twentieth century. This is especially true in the development of factor analysis, where the generally positive intercorrelations among tests and measures of school performance led Spearman (1904) to develop his idea that a latent general factor could explain these correlations. The requirement that a one-factor model could explain all manner of correlations was, of course, too restricted, and a wider range of models for the structure of correlations was introduced by many people, including Hotelling (1933), Thurstone (1935, 1947), and Guttman (1953, 1954). What today we would call exploratory factor analysis is, of course, just one type of hypothesis that can be evaluated by an analysis of correlation coefficients. In a very different but equally brilliant way, Wright (1921, 1934) developed his theory of path analysis to explain correlations based on hypothesized causal influences of variables on each other. See, for examples, Bentler (1986), Bartholomew (2007), and Shipley (2000, Chap. 3) for an exposition of some of the early history, and Cudeck and MacCallum (2007) on some consequences of these early developments.

The range of hypotheses on correlations extends beyond these classical methods and includes such simple illustrative hypotheses that some correlations are zero, or equal; that correlations in repeated measures trail off in a prescribed way; that functions of correlations such as standardized beta coefficients in regression are zero in the population; and so on. Hypotheses on correlation structures are well known (e.g., Brien, James, & Venables, 1988; Brien, Venables, James, & Mayo, 1984; Browne, 1977, 1992; Cheung & Chan, 2004; Fouladi & Steiger, 1999; Jennrich, 1970; Jöreskog, 1978; McDonald, 1975; Olkin & Finn, 1990, 1995; Steiger, 1980a,b), and remain popular today (Preacher, 2006). Such a general class of problems can be approached by hypothesizing that the $p \times p$ population correlation matrix **P** from a set of variables can be expressed in terms of more basic parameters, that is, $\mathbf{P} = \mathbf{P}(\theta)$, where θ is the vector of parameters. In this chapter, we review a variety of approaches for obtaining an estimator $\hat{\theta}$ of θ and the associated standard errors as well as for testing the null hypothesis $\mathbf{P} = \mathbf{P}(\theta)$ based on a sample correlation matrix **R**. The degree of simplicity and practicality of the resulting methods depends on the assumptions that are tenable for the data generation process. Several new methods are proposed that have not yet been studied in depth.

1.2 CORRELATION VERSUS COVARIANCE STRUCTURES

With the excitement generated by Jöreskog's (1967) development of a method to compute the elusive maximum likelihood estimator (MLE) in exploratory factor analysis, focus in the field shifted substantially to the analysis of covariance structures rather than correlations (e.g., Bock & Bargmann, 1966; Browne, 1974; Jöreskog, 1969, 1970; Wiley, Schmidt, & Bramble, 1973). In covariance structure analysis, we hypothesize that the population covariance matrix Σ has the form $\Sigma = \Sigma(\theta)$, where θ is the vector of basic parameters. Jöreskog (1967, 1969) worked with the factor analytic structure $\Sigma = \Lambda \Phi \Lambda' + \Psi$, where Λ is a $p \times k$ matrix of factor loadings, Ψ is a diagonal matrix of unique variances, and Φ is the covariance matrix of the factors taken as $\Phi = \mathbf{I}$ in exploratory factor analysis. In this setup, the basic parameter vector θ consists of the unknown elements in $\{\Lambda, \Phi, \Psi\}$. The class of interesting models was quickly extended to a wider variety of models for the covariance matrix, especially those derived from linear latent variable models such as the Jöreskog–Keesling–Wiley model (Jöreskog, 1977; Wiley, 1973), known as LISREL (Jöreskog & Van Thillo, 1973), the Bentler–Weeks model (Bentler & Weeks, 1980), or the reticular action model (McArdle & McDonald, 1984). For an overview of various meanings of latent variable, see Bollen (2002). We discuss briefly two of these more general types of models and their application to correlation structure analysis.

A LISREL-type model is basically a factor analytic simultaneous equation model. The relationships among the variables in the model can be expressed by two equations, one relating the observed variables to some hypothesized latent variables, and a second describing the relations among latent variables. The measurement model is just the standard factor analysis equation given by

$$\mathbf{x} = \Lambda\xi + \epsilon \tag{1.1}$$

where $\mathbf{x}$ are observed variables, assumed to have means of zero; ξ the latent factors; ϵ unique variates or errors of measurement; and Λ the factor loadings matrix. The simultaneous equation or structural model relating latent variables to each other is given as

$$\xi = \mathbf{B}\xi + \zeta \tag{1.2}$$

where $\mathbf{B}$ is the matrix of regression coefficients for predicting latent variables from latent variables, and ζ are the residuals associated with those regressions. This allows any factor ξ_i to be regressed on any other factor ξ_j.

Assuming no correlations between ξ, ζ, and ϵ, it can be shown that the covariance structure implied by this model is given by

$$\Sigma = \Lambda(\mathbf{I} - \mathbf{B})^{-1}\Phi(\mathbf{I} - \mathbf{B})^{-1'}\Lambda' + \Psi \tag{1.3}$$

where now $\Phi = \text{cov}(\zeta)$ and $\Psi = \text{cov}(\epsilon)$. To apply this model to correlation structures, we note that we can write $\Sigma = \mathbf{D}\mathbf{P}(\theta)\mathbf{D}$, where $\mathbf{D}$ is a diagonal matrix containing standard deviations. Using this form for the covariance matrix allows us to model correlation structures (Jöreskog, 1978), provided that it is possible to enforce the constraint $\text{diag}(\mathbf{P}) = \mathbf{I}$. One way to obtain a model for correlation structures is with

$$\mathbf{P} = \Lambda^*(\mathbf{I} - \mathbf{B})^{-1}\Phi(\mathbf{I} - \mathbf{B})^{-1'}\Lambda^{*'} + \Psi^* \tag{1.4}$$

where $\Lambda^* = \mathbf{D}^{-1}\Lambda$ is a matrix of standardized factor loadings and $\Psi^* = \{\mathbf{I} - \text{diag}(\Lambda^*(\mathbf{I} - \mathbf{B})^{-1}\Phi(\mathbf{I} - \mathbf{B}^{-1'})\Lambda^{*'})\}$. In other words, unique or error variances associated with observed variables are no longer parameters in this correlation structure model. Although technically error covariances can still be parameters, they also can be parameterized in the measurement model by adding a factor. Note that while (1.4) can always be fit as a correlation structure, there are conditions on Λ that would make it impossible to scale to $\Lambda^* = \mathbf{D}^{-1}\Lambda$. For example, if Λ contained fixed nonzero parameters beyond those needed for identification, Λ^* would not maintain those restrictions for an arbitrary $\mathbf{D}$. This is not to say that one cannot estimate Λ^* with many fixed nonzero parameters, but then a covariance structure (1.3) based on $\Lambda = \mathbf{D}\Lambda^*$ would not maintain those restrictions. The issue of the scale invariance of parameters is discussed further below.

The Bentler–Weeks model makes no special distinction between observed and latent variables. The relations among its variates are given by just one equation,

$$\eta = \beta\eta + \gamma\xi \tag{1.5}$$

Here ξ is a vector of independent variables and η is a vector of dependent variables. Independent variables can include observed variables, residuals from latent regressions, measurement errors, latent factors, and so on. The dependent variables can be observed or latent as well. Among the advantages of this approach is that specifications that are not allowed directly in the LISREL model can be handled automatically. For example, an independent observed variable can influence a factor. Let $\nu' = (\eta', \xi')$ represent all the variables, and with $\mathbf{G}$ a known 0–1 matrix, let $\mathbf{x} = \mathbf{G}\nu$ select from all the variables the observed ones. Then with

$$\mathbf{B} = \begin{bmatrix} \beta & 0 \\ 0 & 0 \end{bmatrix} \quad \text{and} \quad \mathbf{\Gamma} = \begin{bmatrix} \gamma \\ \mathbf{I} \end{bmatrix}$$

we have that $\nu = \mathbf{B}\nu + \mathbf{\Gamma}\xi$ and the covariance matrix of the observed variables can be expressed as

$$\mathbf{\Sigma} = \mathbf{G}(\mathbf{I} - \mathbf{B})^{-1}\mathbf{\Gamma}\mathbf{\Phi}\mathbf{\Gamma}'(\mathbf{I} - \mathbf{B})^{-1'}\mathbf{G}' \tag{1.6}$$

where $\mathbf{\Phi} = \text{cov}(\xi)$ is the covariance matrix of independent variables. There are several ways to make this a correlation structure model. The most obvious is to impose the nonlinear constraint $\text{diag}(\mathbf{\Sigma}) = \mathbf{I}$. If the model contains independent error variables that have variances as parameters, then as in (1.4), those elements of $\mathbf{\Phi}$, say $\mathbf{\Phi}_{\epsilon\epsilon}$, are a nonlinear function of the other parameters.

As noted, over the past quarter century, most of the statistical theory for these types of general models has been developed for covariance structures. It is based on the asymptotic distribution of the sample covariance matrix $\mathbf{S}$. For example, the standard approach to (1.3) or to (1.6) is to consider them as covariance structures. Of course, a covariance structure solution is frequently presented in standardized form, where all the parameters have been rescaled such that $\text{diag}(\hat{\mathbf{\Sigma}}) = \mathbf{I}$, because of ease of interpretation. In fact, applied researchers often prefer to work with the correlation matrix $\mathbf{R}$ because of the simple interpretation of correlations and derived quantities such as standardized regression coefficients, as well as the typical arbitrariness of scale of most social science variables. Because appropriate statistical theory is not easily available, a frequent practice is to work with correlation matrices as if they were covariance matrices, that is, to apply covariance structure statistical theory to $\mathbf{R}$. There has been some study of this practice and its consequences (e.g., Cudeck, 1989; Krane & McDonald, 1978; Lee, 1985; Shapiro & Browne, 1990). It turns out that there are situations when this procedure yields at least partially valid statistical inference. If the model is fully scale-invariant (i.e., the model structure is maintained when rescaling $\mathbf{S}$ to yield $\mathbf{R}$), this procedure may yield correct test statistics as well as parameter estimates for the scale-invariant parameters, but it may lead to incorrect standard errors for some parameters. When the model is not fully scale-invariant (e.g., it has some fixed nonzero factor loadings beyond those needed for identification), incorrect parameter estimates, standard errors, and tests will be obtained. The most important source of the problem is that sample-dependent scaling is used to transform $\mathbf{S}$ to $\mathbf{R}$. Sample rather than known population standard deviations are used in this transformation, and this sample dependency is not taken into account when fitting $\mathbf{\Sigma}(\theta)$ to $\mathbf{R}$ using covariance structure statistical methods. As summarized by Cudeck (1989, p. 326), "By analyzing a correlation matrix, one may (a) implicitly alter the model being studied, (b) produce a value of the omnibus test statistic that is incorrect, or (c) report standard errors that are quite discrepant from the correct values". None of these problems occur when correlation structures are studied directly using appropriate statistical theory. In this chapter, we review and extend this theory.

Before we get to correlation structure theory, we do not want to leave the impression that every problem of interest is best studied as a correlation structure problem. Indeed, some problems are handled more appropriately with covariance structure theory and its extensions. Mean structure models provide the most obvious illustration. In such models the variable means μ also have a structure (Sörbom, 1974), so that $\mu = \mu(\theta)$ and $\Sigma = \Sigma(\theta)$, see, e.g., Yuan and Bentler (2006a, b). Perhaps the most popular mean structure models are growth curve models, in which mean trajectories and individual differences in trajectories of scores over time are of primary interest (e.g., Bollen & Curran, 2006; Duncan, Duncan, & Strycker, 2006; Meredith & Tisak, 1990). In such models, in addition to means, variances change systematically. Another example involves the study of possible invariance of covariance structure parameters across populations, as can be found in work on factorial invariance (e.g., Jöreskog, 1971; Millsap & Kwok, 2004).

1.3 ESTIMATION AND MODEL TESTING

1.3.1 Basic Asymptotic Theory

We assume that the observed $N \times p$ data matrix $\mathbf{X}$ is a random sample of size N from a population with the $p \times p$ correlation matrix $\mathbf{P}$, which we hypothesize to be generated according to $\mathbf{P} = \mathbf{P}(\theta)$, where θ is a $q \times 1$ vector of parameters. In all that follows, we assume that this hypothesis is true unless explicitly stated otherwise. Let $n = N - 1$. With continuous data, the sample matrix $\mathbf{R}$ of Pearson correlations is a standard estimator of $\mathbf{P}$, and under normality it is the optimal estimator. In most general terms, estimation of θ is done by minimizing a discrepancy function F, which depends on both the sample correlation matrix $\mathbf{R}$ and on the structured population correlation matrix $\mathbf{P}(\theta)$ proposed. To stress this dependence, we may write $F(\mathbf{r}, \rho(\theta))$, where $\mathbf{r}$ is the $p(p-1)/2 \times 1$ vector of nonredundant sample correlations arranged in a specific way and $\rho(\theta)$ is the corresponding vector of population correlations under the model. For simplicity, let $\tilde{p} = p(p-1)/2$. The following conditions must be true of F: $F(\mathbf{r}, \rho(\theta)) \geq 0$ for all $\mathbf{r}$, $\rho(\theta)$; $F(\mathbf{r}, \rho(\theta)) = 0$ if and only if $\mathbf{r} = \rho(\theta)$; and F is twice continuously differentiable in $\mathbf{r}$ and ρ (Browne, 1982, 1984). Under these very general conditions, the resulting estimator $\hat{\theta} = \arg \min_{\theta} F(\mathbf{r}, \rho(\theta))$ is consistent.

A well-known type of minimum discrepancy function (e.g., Steiger, 1980a,b) is a quadratic form in the residuals:

$$Q_n = (\mathbf{r} - \rho(\theta))' \mathbf{W}_n (\mathbf{r} - \rho(\theta)) \qquad (1.7)$$

Here, $\mathbf{W}_n$ is a $\tilde{p} \times \tilde{p}$ symmetric weight matrix. We use the subscript n to indicate that the weight matrix may not necessarily be a fixed quantity but can instead be estimated from the data. Shapiro (1985) showed that any minimum discrepancy function F is asymptotically a quadratic form with the weight matrix given by $W =$

$\frac{1}{2}(\partial^2 F/\partial r\, \partial r')_{r=\rho}$. This result justifies working with the discrepancy function of the form in (1.7) to derive all necessary asymptotic results.

The optimal choice of $\mathbf{W}_n$ depends on the asymptotic distribution of the vector $\sqrt{n}(\mathbf{r} - \rho)$. From the central limit theorem, we have that

$$\sqrt{n}(\mathbf{r} - \rho) \xrightarrow{d} N(0, \mathbf{V}) \tag{1.8}$$

Here, $\xrightarrow{d}$ indicates convergence in law, and the $\tilde{p} \times \tilde{p}$ matrix $\mathbf{V}$ is the asymptotic covariance matrix of the vector of sample correlations. If for the discrepancy function (1.7) chosen it holds that $\mathbf{W}_n \xrightarrow{p} \mathbf{V}^{-1}$, where $\xrightarrow{p}$ indicates convergence in probability, the estimator obtained by minimizing (1.7) is asymptotically efficient within the class of all quadratic estimators. In this case we say that the weight matrix is "specified correctly," and the resulting estimator is optimal in the sense that it will have the smallest asymptotic variance. In particular, defining $\tilde{p} \times q$ matrix of model derivatives $\dot{\rho} = \partial \rho(\theta)/\partial \theta'$, it can be shown that when the weight matrix is specified correctly,

$$\sqrt{n}(\hat{\theta} - \theta) \xrightarrow{d} N(0, (\dot{\rho}'\mathbf{V}^{-1}\dot{\rho})^{-1}) \tag{1.9}$$

Furthermore, the quantity

$$T = n\hat{Q}_n = nQ_n(\hat{\theta}) \tag{1.10}$$

is asymptotically chi-square distributed with $d = \tilde{p} - q$ degrees of freedom. Thus, T can be used as a test statistic to evaluate the null hypothesis $H_0 = \mathbf{P} = \mathbf{P}(\theta)$. Provided that one has an appropriate $\mathbf{W}_n$, this minimum χ^2 theory (Ferguson, 1958, 1996) is all the theory needed to estimate the parameters of the model, obtain standard errors, and evaluate the model's adequacy. Appropriate adjustments can be made to (1.9) and the degrees of freedom associated with (1.10) when estimating the model under constraints such as $h(\theta) = 0$ (see, e.g., Bentler & Dijkstra, 1985; Lee & Bentler, 1980; Mels, 2000). We do not discuss such constraints further.

1.3.2 Distribution of T Under Model Misspecification

Of course, models may be incorrect; that is, the null hypothesis H_0: $\mathbf{P} = \mathbf{P}(\theta)$ may not be true. The distribution of T under the alternative hypothesis H_1 is worth noting. In this case, as N increases, T tends to infinity. To obtain an asymptotic distribution, we make the typical assumption of a sequence of local alternatives (also known as parameter drift). That is, we assume that the true population covariance matrix $\mathbf{P}_N$ depends on N and converges to $\mathbf{P}(\theta)$ at the rate of $1/\sqrt{N}$. An informal way of stating this is that model misspecification is "not too large" and is within the sampling error for any given N. Then the asymptotic distribution of T is noncentral chi-square with d degrees of freedom and noncentrality parameter $\delta = \lim_{N\to\infty} N(\rho_N - \rho(\hat{\theta}))'\mathbf{V}^{-1}(\rho_N - \rho(\hat{\theta}))$. This equation makes clear that without the parameter drift assumption, noncentrality will increase with N, resulting in no asymptotic distribution for T. A sample estimator can be obtained as

$\hat{\delta} = T - d$; see Raykov (2005) for an alternative estimator. Although this noncentral chi-square approximation is typically used in studies of power in covariance structure analysis (Curran et al., 2002; MacCallum, Browne, & Sugawara, 1996; Satorra, 2003; Satorra & Saris, 1985) and to define "goodness-of-fit" indices such as the comparative fit index, root mean square error of approximation, and others (e.g., Bentler, 1990; Browne & Cudeck, 1993; Curran et al., 2003; Kim, 2005; McDonald & Marsh, 1990; Steiger & Lind, 1980), its use has been criticized by several authors (e.g., Li & Bentler, 2006; Yuan, 2005; Yuan, Hayashi, & Bentler, 2007; Yuan & Marshall, 2004). The approximation can be expected to be particularly bad when $\rho - \rho(\hat{\theta})$ is large and also when $\mathbf{W}_n$ depends on the model (as will be the case with normal theory ML estimation), because in that case $\mathbf{W}_n$ is no longer consistent for $\mathbf{V}^{-1}$ without the parameter drift assumption. Since one never knows a priori how serious a misspecification actually may be, it seems prudent to remember that the noncentral chi-square approximation may be seriously compromised if $\mathbf{W}_n$ is way off as an estimate of $\mathbf{V}^{-1}$. We do not discuss model misspecification further.

1.3.3 Distribution of T Under Weight Matrix Misspecification

The asymptotic covariance matrix of sample correlations $\mathbf{V}$ depends on the distribution of the data, which may be well behaved and known (e.g., multivariate normal), or difficult or impossible to specify. Our choice of $\mathbf{W}_n$ for use in (1.7) may therefore be wrong in the sense that $\mathbf{W}_n \xrightarrow{P} \mathbf{V}^{-1}$ may not hold. The resulting estimator is still consistent, but it will no longer be optimal. Assuming that $\mathbf{W}_n \xrightarrow{P} \mathbf{W}$, it can be shown that under $\mathbf{P} = \mathbf{P}(\theta)$,

$$\sqrt{n}(\hat{\theta} - \theta) \xrightarrow{d} N(0, (\dot{\rho}'\mathbf{W}\dot{\rho})^{-1}\dot{\rho}'\mathbf{W}\mathbf{V}\mathbf{W}\dot{\rho}(\dot{\rho}'\mathbf{W}\dot{\rho})^{-1}) \qquad (1.11)$$

(e.g., Bentler & Dijkstra, 1985; Browne, 1984). The covariance matrix in (1.11) yields larger standard errors than those obtained from (1.9), although (1.11) reduces to (1.9) when $\mathbf{W} = \mathbf{V}^{-1}$ (i.e., when the weight matrix is specified correctly). Furthermore, the quantity $T = n\hat{Q}_n$ in (1.10) is generally no longer asymptotically chi-square distributed (for exceptions, see the literature on asymptotic robustness; e.g., Amemiya & Anderson, 1990; Browne, 1987; Browne & Shapiro, 1988; Satorra, 2002). As noted by Bentler and Dijkstra (1985), Satorra and Bentler (1986, 1994), and others, the asymptotic distribution of T can instead be characterized as a weighted mixture of d independent chi-square variates with one degree of freedom. The weights $w_1, \ldots, w_d$ are the nonzero eigenvalues of the matrix product $\mathbf{UV}$, where $\mathbf{U} = \mathbf{W} - \mathbf{W}\dot{\rho}(\dot{\rho}'\mathbf{W}\dot{\rho})^{-1}\dot{\rho}'\mathbf{W}$. That is, asymptotically,

$$T \sim \sum_{i=1}^{d} w_i \chi_1^2 \qquad (1.12)$$

When $\mathbf{W}_n$ is specified correctly, all w_i's are 1 and the distribution in (1.12) simplifies to χ_d^2, as noted for (1.10). As we will see later, the distribution of T under distributional misspecification is useful for deriving additional tests of model fit when the weight matrix is misspecified.

1.3.4 Estimation and Testing with Arbitrary Distributions

We now discuss specific estimation and testing methods. The most versatile approach, at least in theory, is to make no assumptions about the distribution of the p continuous variables. A completely unstructured estimate of the matrix V in (1.9) is obtained from the data, and estimation is done by setting $W_n = \hat{V}^{-1}$ and minimizing (1.7). This approach is known as the generalized least squares (Dijkstra, 1981), minimum distance (Chamberlain, 1982; Satorra, 2001), asymptotically distribution free (ADF; Browne, 1982, 1984), or minimum χ^2 method (Ferguson, 1958). To implement this approach we need the general form of the asymptotic covariance matrix of sample correlation coefficients. This matrix has been known for a long time, and rediscovered several times (e.g., Browne & Shapiro, 1986; de Leeuw, 1983; Hsu, 1949; Neudecker & Wesselman, 1990; Steiger & Hakstian, 1982). We first define a general fourth-order moment as

$$\sigma_{ijkl} = E(x_i - \mu_i)(x_j - \mu_j)(x_k - \mu_k)(x_l - \mu_l) \qquad (1.13)$$

where $1 \leq i, j, k, l \leq p$, x_i indicates the ith variable, and $\mu_i = E(x_i)$. For reference we note that (1.13) is used along with population covariances $\sigma_{ij} = E(x_i - \mu_i)(x_j - \mu_j)$ to describe the asymptotic covariance of sample covariances s_{ij} as

$$\text{acov}(\sqrt{n}s_{ij}, \sqrt{n}s_{kl}) = \sigma_{ijkl} - \sigma_{ij}\sigma_{kl} \qquad (1.14)$$

The asymptotic covariance of sample correlations r_{ij} is a bit more complex. Using (1.13), we define a standardized fourth-order moment as

$$\rho_{ijkl} = \sigma_{ijkl}/(\sigma_{ii}\sigma_{jj}\sigma_{kk}\sigma_{ll})^{1/2} \qquad (1.15)$$

where $\sigma_{ii} = E(x_i - \mu_i)^2$. Let ρ_{ij} be a typical element of the population correlation matrix P. Then a typical element of V, $v_{ij,kl} = \text{acov}(\sqrt{n}r_{ij}, \sqrt{n}r_{kl})$, is given by

$$v_{ij,kl(ADF)} = \rho_{ijkl} + .25\rho_{ij}\rho_{kl}(\rho_{iikk} + \rho_{jjkk} + \rho_{iill} + \rho_{jjll})$$
$$- .5\rho_{ij}(\rho_{iikl} + \rho_{jjkl}) - .5\rho_{kl}(\rho_{ijkk} + \rho_{ijll}) \ . \qquad (1.16)$$

One approach to computing the ADF estimate of V is by substituting sample quantities into (1.16). That is,

$$\hat{v}_{ij,kl(ADF)} = r_{ijkl} + .25r_{ij}r_{kl}(r_{iikk} + r_{jjkk} + r_{iill} + r_{jjll})$$
$$- .5r_{ij}(r_{iikl} + r_{jjkl}) - .5r_{kl}(r_{ijkk} + r_{ijll}) \qquad (1.17)$$

Here, the r_{ij}'s are sample correlations and the r_{ijkl}'s are standardized sample fourth order moments. An efficient way of computing this estimate $\hat{V}_{ADF}$ is given by

Mooijaart (1985). Because the ADF methodology makes no assumptions about the distribution of the data other than that V is finite, the weight matrix $\mathbf{W}_n = \hat{\mathbf{V}}_{ADF}^{-1}$ based on (1.17) is always specified correctly, the resulting estimator is asymptotically efficient, and the resulting test statistic T_{ADF} is asymptotically chi-square distributed.

Although the ADF approach to correlation structures has attractive asymptotic properties, there is reason to be cautious, since the method requires estimating $\tilde{p}(\tilde{p}+1)/2$ distinct parameters (elements of V) from the data available. If the model has $p = 10$ variables, there are $\tilde{p} = 45$ distinct elements in the correlation matrix, and hence 1035 unique elements in V. If the model has 30 variables, the number of unique elements in V increases to 94,830. A very large sample size would be required to estimate such a big matrix accurately when no structure is imposed on it. Furthermore, to perform ADF estimation, we require not $\hat{\mathbf{V}}_{ADF}$ itself but its inverse, and inverting such a large matrix can further distort the results or lead to numerical problems. In the context of covariance structure analysis, the test statistic T_{ADF} tends to exhibit inflated rejection rates unless the sample size is in the thousands or the number of variables is small (e.g., Chou, Bentler, & Satorra, 1991; Curran, West, & Finch, 1996; Hu, Bentler, & Kano, 1992). Thus, there are good reasons to believe that the ADF approach with correlation structures will also not be very stable under similar conditions. To our knowledge, only one study evaluated the performance of the ADF approach with correlation structures, and found that in fact this method performed even worse than the ADF approach with covariance structures (Mels, 2000). This may be because equation (1.16) is much more complicated than equation (1.14) and relies on accurate estimation of multiple fourth-order moments. We present the results of a small simulation in Section 1.5.

One solution to this problem is to employ two-stage ADF estimation, which we refer to as structured ADF (Mels, 2000; Steiger, 2005; Steiger & Hakstian, 1982). In the first stage, model-based estimates of sample correlations are obtained using a simpler method, such as least squares (LS), generalized least squares (GLS), or normal theory reweighted least squares (RLS), which are defined later in this chapter. In the second stage, these estimates are used instead of sample correlations in equation (1.17) to compute a "structured" or "two-stage" distribution-free estimate of V, say $\hat{\mathbf{V}}_{TADF}$. ADF estimation is then conducted with a structured weight matrix. To our knowledge, only two studies evaluated the performance of this methodology (Fouladi, 2000; Mels, 2000), and both studied the structured ADF approach with LS estimation as the first stage. That is, model-based estimates ρ_{ij} were obtained by minimizing equation (1.7) with $\mathbf{W}_n = \mathbf{I}$. The structured ADF estimate of V was computed as

$$\hat{v}_{ij,kl(TADF)} = r_{ijkl} + .25\hat{\rho}_{ij}\hat{\rho}_{kl}(r_{iikk} + r_{jjkk} + r_{iill} + r_{jjll})$$
$$- .5\hat{\rho}_{ij}(r_{iikl} + r_{jjkl}) - .5\hat{\rho}_{kl}(r_{ijkk} + r_{ijll}) . \quad (1.18)$$

The weight matrix was then set to be the inverse of $\hat{\mathbf{V}}_{TADF}$. Mels (2000) found that the structured ADF estimator was less biased than the original ADF estimator based on (1.17). The structured ADF test statistic had a distribution that resembled chi-square much more closely than the original ADF test statistic. However, both approaches underestimated the standard errors and the "structured" approach to a greater extent. Fouladi (2000) further found that the structured ADF test statistic performed well but tended to be conservative under some conditions, leading to rejection rates that are too low. We do a small simulation study of the reweighted least squares (RLS) version of the structured ADF approach in Section 1.5, and we propose other uses for the structured estimate of V in the next section. Finally, we note that it is possible to compute a structured distribution-free estimate of V in the context of covariance structures as well, by using model-reproduced covariances in the second term of (1.14). Surprisingly, however, we have found that this modification makes almost no difference (see also Tanaka & Bentler, 1985), and hence we do not consider it when comparing correlation and covariance structure methods.

A different approach is to accept the original unstructured ADF parameter estimates but to modify the ADF test statistic so that it is better behaved. Several improvements to the ADF statistic have been proposed in the context of covariance structures. Yuan and Bentler (1997a) proposed computing a new test statistic as follows. Once the model has been estimated and the ADF model–reproduced correlations have been obtained, use them in place of sample covariances σ_{ij} in (1.14) to compute the test statistic. This approach may appear similar to the structured two-stage ADF approach; however, there are two major differences. The structured matrix $\hat{V}_{ADF}$ uses ADF estimates of correlations and not LS or some other initial estimates, and the structured matrix is used only to compute the test statistic, not to obtain the estimates themselves. The main appeal of the Yuan–Bentler statistic is that it is easily computed; it actually turns out to be a simple rescaling of the original ADF statistic: $T_{YB(ADF)} = T_{ADF}/(1+T_{ADF}/n)$. As the sample size gets large, $T_{YB(ADF)}$ becomes equal to T_{ADF}, but at smaller sample sizes it can help to eliminate the typical inflation of the ADF statistic. Unfortunately, the same does not hold for correlation structures: while the Yuan–Bentler statistic can be proposed in theory as the statistic that uses the structured estimator of (1.16), it no longer is a simple rescaling of the ADF statistic for correlation structures.[1] It still may be worthwhile to compute and study this statistic, however. We expect its empirical performance to be similar to its covariance structure counterpart.

A second improvement of the covariance structure ADF test statistic was developed by Yuan and Bentler (1999) and is given by

$$T_{F(ADF)} = \frac{N - (\tilde{p} - q)}{n(\tilde{p} - q)} T_{ADF} \qquad (1.19)$$

---

[1] Another justification for the Yuan–Bentler correction in covariance structures is that the ADF matrix in (1.14) can also be written as a covariance matrix of $y_t = \text{vech}(x_t - \bar{x})(x_t - \bar{x})'$, where $\bar{y} = s$, the vector of sample covariances. The Yuan–Bentler statistic then simply computes the covariance matrix using $\hat{\sigma}$ in place of $\bar{y}$. A parallel extension exists for correlation structures. The matrix in (1.16) can be defined as a covariance matrix of $m_t = \text{vec}(z_t z_t') - .5\text{vec}\{R \odot (1z_t^{2'} + z_t^2 1')\}$ where $\odot$ is the element-wise (Hadamard) product. In this case, however, $\bar{m}$ is identically zero and cannot be structured. Under this justification, an extension of the Yuan–Bentler statistic to correlation structures simply does not exist.

The distribution of this statistic is approximated by an F distribution with $d = (\tilde{p} - q)$ and $N - d$ degrees of freedom. The rationale here is the analogy with Hotelling's T^2 statistic in multivariate analysis (e.g., Seber, 2004), which is used to test hypotheses about the mean vector when the data come from a multivariate normal distribution with unknown covariance matrix. The analogy is not complete: the vector $\sqrt{n}(\mathbf{r} - \rho)$ is not normal for all n but only approaches normality asymptotically, and $\rho(\theta)$ is not a linear function. The F distribution is only an approximation to the small-sample behavior of $T_{F(ADF)}$; however, it is asymptotically correct. The asymptotic distribution of (1.19) is an F distribution with d and ∞ degrees of freedom, which is a scaled chi-square variate, $1/d \, \chi_d^2$. Thus, even though the statistic in (1.19) is not asymptotically equivalent to T_{ADF}, they result in equivalent tests, in the sense that they will tend to produce identical p-values as N gets large. In the context of covariance structure analysis, $T_{F(ADF)}$ has been found to perform extremely well; however, the simulations have been limited. This statistic has never been applied to correlation structure analysis, but the extension is straightforward in this case, and may result in similar improvement in performance.

The unstructured asymptotic covariance matrix estimator $\hat{\mathbf{V}}_{ADF}$ given by (1.16) can be used in (1.9) to obtain estimates of the standard errors of the parameter estimates. However, in the context of covariance structure analysis, Yuan and Bentler (1997b) found that these estimates did not behave as well as desired. They proposed to correct the covariance matrix in (1.9) with a scalar multiplier that in our context is

$$\frac{n}{n - \tilde{p} - 1}(\hat{\rho}' \hat{\mathbf{V}}_{ADF}^{-1} \hat{\rho})^{-1} \tag{1.20}$$

With covariance structures, Yuan and Bentler (1997b) found that this estimator worked better in practice. The justification for this correction again comes from standard multivariate analysis. With multivariate normal data, the inverse of the multiplier in (1.20) is the adjustment necessary to obtain an unbiased estimate of the inverse of the sample covariance matrix. Since $\hat{\mathbf{V}}_{ADF}$ can be viewed as a covariance matrix of some transformation of the data (see footnote 1) for both covariances and correlations, we propose that the modified standard errors in (1.20) may also be useful in correlation structure analysis.

In summary, ADF estimation is a method with very attractive theoretical properties, but it may be computationally and statistically problematic in small samples or with large models. In later sections we review some alternatives. All of them make restrictive assumptions about $\mathbf{V}$ and define $\mathbf{W}_n$ accordingly. When the assumptions are met and $\mathbf{W}_n$ is specified correctly, estimation is greatly stabilized. However, it is possible that $\mathbf{W}_n$ may be specified incorrectly. Before we address specific choices for $\mathbf{W}_n$, we first discuss some model tests of fit that still yield accurate conclusions under weight matrix misspecification. We discuss the following three types of tests: tests that are based on the mixture distribution, residual-based tests and their modifications, and scaled tests. All these tests require a consistent estimate of $\mathbf{V}$. This estimate can be obtained by using either the original ADF estimator given by (1.17) or a structured version given by (1.18). Thus, the ADF methodology has applications that extend beyond the classical distribution-free estimation approach presented in this section.

1.3.5 Tests of Model Fit Under Distributional Misspecification

1.3.5.1 Tests Based on a Mixture of χ_1^2.

Suppose first that the weight matrix is possibly (or surely) misspecified. Then an obvious thing to do is to refer the statistic T to its true distribution as given in (1.12): namely, $T \sim \sum_{i=1}^{d} w_i \chi_1^2$, where $w_1, \ldots, w_d$ are the nonzero eigenvalues of the matrix product $\mathbf{UV}$, where $\mathbf{U} = \mathbf{W} - \mathbf{W}\dot{\rho}(\dot{\rho}'\mathbf{W}\dot{\rho})^{-1}\dot{\rho}'\mathbf{W}$. These weights are not known and will need to be estimated from the data. We can easily obtain the estimator $\hat{\mathbf{U}}$ by using sample quantities in the definition of $\mathbf{U}$. However, the weights also depend on the correct asymptotic covariance matrix $\mathbf{V}$. One approach is to use the distribution-free estimator $\hat{\mathbf{V}}_{ADF}$ of $\mathbf{V}$ in (1.17). Then the estimated weights $\hat{w}_i$ can be obtained from the nonzero eigenvalues of the matrix product $\hat{\mathbf{U}}\hat{\mathbf{V}}$, and we may compute the probability

$$\text{pr}\left(\sum_{i=1}^{d} \hat{w}_i \chi_1^2 < t\right) \tag{1.21}$$

to evaluate the null hypothesis using some cutoff value t, which is the critical value past which, say, 5% of the mixture distribution is contained. Finding t for a given set of estimated weights is also not an easy matter and may require special programming in a package such as Matlab or R. Despite these difficulties, this test is appealing because it makes use of the actual distribution of T under model misspecification, instead of making approximations.

Very little empirical work has been done on the mixture test, however. Bentler (1994) proposed and studied this test with covariance structure analysis, using an algorithm of Gabler and Wolff (1987) to compute the given probability. He found the method to work quite well, except under conditions of dependence of observations and at small sample sizes. We expect that problems with the method are due to poor estimation of the eigenvalues $\hat{w}_i$. Nothing is known about the behavior of these eigenvalues in correlation structures. On the one hand, they may be expected to be better behaved because correlations have a restricted range. On the other hand, they may share the fate of the correlation structure ADF methodology, which does even worse than the corresponding covariance structure ADF methodology. Modifications, are, however, possible. The mixture test can be implemented using shrunken or trimmed estimators of the eigenvalues $\hat{w}_i$, and possibly using alternative computational methods for obtaining the required probabilities (Davies, 1980; Farebrother, 1984; Wood, 1989). Additionally, using the structured ADF estimator $\hat{\mathbf{V}}_{TADF}$ given by (1.18) may improve performance. In sum, using (1.21) directly to evaluate the model test of fit in correlation structures remains an open research problem.

1.3.5.2 Tests Based on Residuals.

Actually, there are some simpler tests than the mixture test given by (1.21). In fact, there is a statistic that behaves asymptotically as a chi-square variate under the null hypothesis regardless of whether the weight matrix used in estimation is misspecified. This is advantageous, as using a chi-square distribution is simpler than evaluating a mixture such as (1.21). This statistic was first proposed by Browne (1984) for covariance structure analysis and is referred to as a residual-based statistic. For correlation structures, the residual-based statistic is given by

$$T_{RES} = n(\mathbf{r} - \hat{\rho})' \hat{\mathbf{U}}_V (\mathbf{r} - \hat{\rho}) \qquad (1.22)$$

where $\hat{\rho} = \rho(\hat{\theta})$, $\hat{\theta}$ is obtained by minimizing (1.7) for some $\mathbf{W}_n$, and $\hat{\mathbf{U}}_V$ is a consistent estimator of $\mathbf{U}_V = \mathbf{V}^{-1} - \mathbf{V}^{-1}\dot{\rho}(\dot{\rho}'\mathbf{V}^{-1}\dot{\rho})^{-1}\dot{\rho}'\mathbf{V}^{-1}$. This statistic has an asymptotic chi-square distribution with $d = \tilde{p} - q$ degrees of freedom even if $\mathbf{W}_n \xrightarrow{p} \mathbf{W} \neq \mathbf{V}^{-1}$. In practice, to obtain a consistent estimate of $\mathbf{U}_V$, we need a consistent estimate of $\mathbf{V}$ (or its inverse). The ADF estimator $\hat{\mathbf{V}}_{ADF}$ given by (1.17) can be used. The structured two-stage ADF estimator $\hat{\mathbf{V}}_{TADF}$ given in (1.18) using the same $\hat{\rho}$ from the residuals in the first stage can also be used. In later sections we introduce other forms of residual-based statistics which use estimators of $\mathbf{V}$ that require additional assumptions about the distribution of the data. When these assumptions hold, the resulting test statistic will be better behaved. Thus, (1.22) actually refers to a class of test statistics.

In covariance structure analysis, the residual-based statistic in (1.22) exhibits small sample properties similar to those of the original ADF statistic, for similar reasons. The ADF estimator $\hat{\mathbf{V}}_{ADF}$ requires estimating a lot of parameters and hence may be unstable in small sample sizes. Inverting the resulting matrix may be difficult to do accurately in small samples or with a large number of variables. Worse yet, the matrix $\hat{\mathbf{V}}_{ADF}$ may turn out to be singular or near singular, leading the residual-based test to break down. One way around this problem is to use a generalized inverse of $\hat{\mathbf{V}}$ instead of the regular inverse $\hat{\mathbf{V}}^{-1}$; however, little is known about the effectiveness of this modification. As far as we know, the class of residual-based statistics for correlation structures has not been studied. We can expect, however, that the correlation structure version of this statistic based on $\hat{\mathbf{V}}_{ADF}$ will perform poorly, because its performance should be similar to the performance of the ADF statistic itself. The residual-based statistic based on the structured ADF estimator $\hat{\mathbf{V}}_{TADF}$ may do better. We report on a small simulation study in Section 1.5.

For covariance structures, Yuan and Bentler (1998) proposed corrections to the residual-based statistic designed to improve its small-sample performance. These modified statistics are very similar to the modified ADF statistics given above, and their rationale is exactly the same. We suggest using, parallel to (1.19),

$$T_{F(RES)} = \frac{N - (\tilde{p} - q)}{n(\tilde{p} - q)} T_{RES} . \qquad (1.23)$$

As before, $T_{F(RES)}$ is referred to an F distribution with d and $N - d$ degrees of freedom and is distributed asymptotically as $1/d\,\chi_d^2$. The performance of this new test in the context of correlation structure analysis remains to be established.

1.3.6 Scaled and Adjusted Statistics

An entirely different approach to evaluating model fit in the case of distributional misspecification was proposed by Satorra and Bentler (1988, 1994) in the context of covariance structure analysis, although the idea of scaling a statistic to improve its performance goes back to Bartlett (1950). They proposed to scale the statistic in (1.12), generally distributed as a mixture, so that its expectation is equal to that of a chi-squared variate with d degrees of freedom. For reference we note that if $x \sim \chi_d^2$, then $\mathbf{E}(x) = d$ and $\mathrm{var}(x) = 2d$. We can compute the expected value of (1.12) as $\mathbf{E}(T) = \sum_{i=1}^d w_i = \mathrm{tr}(\mathbf{U}\mathbf{V})$, since the components of the mixture are independent. The scaled statistic is defined as

$$\bar{T}_1 = \frac{d}{\mathrm{tr}(\hat{\mathbf{U}}\hat{\mathbf{V}})}T \tag{1.24}$$

It is clear that $\bar{T}_1 < T$ if $\mathrm{tr}(\hat{\mathbf{U}}\hat{\mathbf{V}}) > d$, thus reducing an inflated T. This is a standard situation in covariance structure analysis, where a weight matrix that is misspecified due to distributional violations tends to result in inflated tests T. The new statistic $\bar{T}_1$ is generally not χ^2 distributed, but its asymptotic expectation is d. Thus, referring it to a χ^2 distribution can be a good approximation, especially when the coefficient of variation of the weights w_i is small (Yuan & Bentler, 1998). When the weight matrix is specified correctly, all the weights w_i are 1 in the population, and the correction has no effect asymptotically. Furthermore, the correction yields an asymptotic χ^2 variate in some special cases: for example, when the distribution of the data is elliptical (as defined below). As before, this statistic requires that an estimate of $\mathbf{V}$ be provided, but unlike with the residual-based statistic, here it does not need to be inverted, which may aid in stability. The tradition is to again use the ADF estimator $\hat{\mathbf{V}}$, although in principle, assumptions can be made that would lead to alternative corrections. $\bar{T}_1$ is most commonly known as the Satorra–Bentler scaled chi-square and is well studied in covariance structure analysis. The version of the Satorra–Bentler chi-square where T is the normal theory maximum likelihood statistic has been found to perform well under a wide variety of nonnormal conditions (e.g., Chou & Bentler, 1995; Chou, Bentler, & Satorra, 1991; Curran, West, & Finch, 1996; Fouladi, 2000; Hoogland, 1999; Hu, Bentler, & Kano, 1992; Nevitt, 2000). With an additional Bartlett-type multiplicative correction factor of $\{1 - [(2p + 4k + 5)/6n]\}$, where k is the number of factors as proposed by Fouladi (2000), it was found to be the best performing of various statistics studied by Nevitt and Hancock (2004). The original (1.24) or its Bartlett modification has not been evaluated in the context of correlation structures. With correlation structures, another version of the Satorra–Bentler chi-square that utilizes the structured two-stage estimator $\hat{\mathbf{V}}_{TADF}$, where the first-stage estimates are those used in the computation of T, is also possible.

Satorra and Bentler (1988, 1994) also proposed a mean- and variance-adjusted statistic, which rescales (1.12) in a way as to make its mean and variance equal to those of a chi-square variate with k degrees of freedom, for some k. That is, we want to find a scaling constant c such that $2\mathbf{E}(cT) = \mathrm{var}(cT)$. Again using independence of the mixture components, we have that $E(cT) = c\sum_{i=1}^{d} w_i = c\,\mathrm{tr}(\mathbf{UV})$ and $\mathrm{var}(cT) = 2c^2 \sum_{i=1}^{d} w_i^2 = 2c^2\mathrm{tr}(\mathbf{UVUV})$. This yields $c = \mathrm{tr}(\mathbf{UV})/\mathrm{tr}(\mathbf{UVUV})$. The mean- and variance-adjusted test statistic is then

$$\bar{T}_2 = \hat{c}T \tag{1.25}$$

where $\hat{c}$ is a consistent estimate of c, with approximate degrees of freedom given by $\hat{k}$, a consistent estimate of $k = E(cT) = [\mathrm{tr}(\mathbf{UV})]^2/\mathrm{tr}(\mathbf{UVUV})$, possibly rounded to its nearest integer. This statistic, referred to as the Satorra–Bentler mean- and variance-adjusted statistic, has been found to perform well in the context of covariance structure analysis. Some authors report it outperforming the scaled statistic $\bar{T}_1$ (Fouladi, 2000), but it has been faulted for low power (Nevitt & Hancock, 2004). It has not been studied in the context of correlation structures for continuous data.

In summary, many options exist for evaluating model fit under distributional misspecification. We have not covered all the possible options, such as the bootstrap-based tests (Bollen & Stine, 1993; Enders, 2002; Yuan & Hayashi, 2003). We also note that even though we do not discuss standard errors in detail, whenever misspecification is suspected, robust standard errors obtained from the covariance matrix in (1.11) should be used, regardless of how the model fit is evaluated. In the next few sections, we return to the assumption that the model is specified correctly, and describe several specific estimation and testing approaches. These approaches all first make some assumptions about $\mathbf{V}$, and then choose $\mathbf{W}_n$ accordingly.

1.3.7 Normal Theory Estimation and Testing

The multivariate normal distribution has a density proportional to

$$|\Sigma|^{-1/2} \exp\left[-\frac{1}{2}(\mathbf{x} - \mu)'\Sigma^{-1}(\mathbf{x} - \mu)\right] \tag{1.26}$$

It can be shown that when the data follow this distribution, the general fourth-order moment given in (1.14) can be expressed as a function of only second-order moments (variances and covariances) as follows

$$\sigma_{ijkl} = \sigma_{ij}\sigma_{kl} + \sigma_{ik}\sigma_{jl} + \sigma_{il}\sigma_{jk} \tag{1.27}$$

where $\sigma_{ij} = E(x_i - \mu_i)(x_j - \mu_j)$, as before. Then the standardized fourth-order moment in (1.15) simplifies to

$$\rho_{ijkl} = \rho_{ij}\rho_{kl} + \rho_{ik}\rho_{jl} + \rho_{il}\rho_{jk} \tag{1.28}$$

Substituting (1.28) into the general expression for an element $v_{ij,kl}$ of $\mathbf{V}$ in (1.16), we obtain that under normal theory, the asymptotic covariance matrix of sample correlations has the following structure (Olkin & Siotani, 1976; Steiger & Hakstian,

1982):

$$v_{ij,kl(NT)} = .5\rho_{ij}\rho_{kl}(\rho_{ik}^2 + \rho_{il}^2 + \rho_{jk}^2 + \rho_{jl}^2) + \rho_{ik}\rho_{jl} + \rho_{il}\rho_{jk}$$
$$- \rho_{ij}(\rho_{jk}\rho_{jl} + \rho_{ik}\rho_{il}) - \rho_{kl}(\rho_{jk}\rho_{ik} + \rho_{jl}\rho_{il}) \quad (1.29)$$

It should be clear from this equation that the advantage of assuming a parametric distribution for the data is that it structures the asymptotic covariance matrix $\mathbf{V}$, making its estimation a much easier task. The expression in (1.29), say $\mathbf{V}_{NT}$, depends only on the population correlations, that is, on a total of $\tilde{p}$ parameters. With 30 variables, we only need to estimate $30(29)/2 = 435$ parameters to obtain an estimate of $\mathbf{V}$ under the assumption of normality, compared to 94,830 needed for the ADF estimator.

To minimize (1.7) we would have to invert the $\tilde{p} \times \tilde{p}$ estimate of (1.29), which requires a lot of computational effort. Although this can certainly be done, Jennrich (1970) proposed a much more efficient way of proceeding. He derived the symbolic form of the inverse of $\mathbf{V}$ under normal theory. The typical element of $\mathbf{V}^{-1}$ when $\mathbf{V}$ has the structure given in (1.29) is

$$\mathbf{V}_{ij,kl(NT)}^{-1} = \rho^{ik}\rho^{jl} + \rho^{il}\rho^{jk} - \rho^{ij}\rho^{kl}(\pi^{ik} + \pi^{jk} + \pi^{il} + \pi^{jl}) \quad (1.30)$$

where $(P^{-1})_{ij} = \rho^{ij}$; that is, ρ_{ij} is the typical element of P^{-1}, $(\Pi^{-1})_{ij} = \pi^{ij}$, $(\Pi)_{ij} = \pi_{ij} = \delta_{ij} + \rho_{ij}\rho^{ij}$, and δ_{ij} is Kronecker's delta. To compute (1.30), we only need to invert $p \times p$ matrices, which greatly simplifies the computations. In fact, we do not need to compute the function (1.7) directly either. Jennrich (1970) gave a simplified expression for the quadratic discrepancy function in (1.7) based on the inverse in (1.30):

$$Q_{n(NT)} = \text{tr}[(\mathbf{R} - \mathbf{P}(\theta))\mathbf{P}_0^{-1}]^2/2 - \text{diag}[(\mathbf{R} - \mathbf{P}(\theta))\mathbf{P}_0^{-1}]'$$
$$\times (\mathbf{I} + \mathbf{P}_0 \odot \mathbf{P}_0^{-1})^{-1} \text{diag}[(\mathbf{R} - \mathbf{P}(\theta))\mathbf{P}_0^{-1}] \quad (1.31)$$

where $\odot$ refers to Hadamard (elementwise) product and $\mathbf{P}_0$ is the true population correlation matrix. This expression is also given by Browne (1977). In practice, minimizing (1.31) directly is not possible because $\mathbf{P}_0$ is not known. We now have several possibilities. As noted by Shapiro and Browne (1990), any consistent estimator of $\mathbf{P}_0$ can be used in (1.31) and hence $n\hat{Q}_{n(N)}$ will be an asymptotic χ_d^2 variate and the estimator will be asymptotically efficient. In particular, they proposed to use $\mathbf{P} = \mathbf{P}(\theta)$, which uses $\partial\mathbf{P}_0/\partial\theta = \partial\mathbf{P}(\theta)/\partial\theta$ in defining the gradient and approximate Hessian for the optimization and yields an asymptotically efficient estimator related to Swain's (1975) "F_1" covariance structure estimator. Let us call this a weighted least squares estimator (WLS). Bentler (2002–2007, the EQS 6 program) proposed to consider $\mathbf{P}_0$ as a fixed weight matrix such that $\partial\mathbf{P}_0/\partial\theta$. The choices for such weight matrices are (1) $\mathbf{P}_0 = \mathbf{R}$, the sample correlation matrix, yielding a generalized least squares (GLS) method; (2) $\mathbf{P}_0 = \mathbf{P}(\hat{\theta}^{(k)})$, the current iteration's model-based estimate, yielding an iteratively reweighted least squares (RLS) method (see Lee & Jennrich, 1979 on how RLS yields MLE in covariance structures); or (3) $\mathbf{P_0} = \mathbf{I}$, which yields a least squares (LS) method, discussed later. For completeness, we may also propose some new diagonally weighted methods related to the WLS, GLS, and RLS estimators such as $\hat{\mathbf{P}}_0^{-1} = \text{diag}(\mathbf{P}(\theta)^{-1})$, $\hat{\mathbf{P}}_0^{-1} = \text{diag}(\mathbf{R}^{-1})$, or $\hat{\mathbf{P}}_0^{-1} = \text{diag}(\mathbf{P}(\hat{\theta}^{(k)})^{-1})$, which, in really huge problems, may be a useful compromise between optimally weighted and LS estimates.

The WLS, GLS, and RLS methods all yield asymptotically efficient estimates and their associated test statistics $n\hat{Q}_{n(NT)}$ are distributed asymptotically as χ_d^2. Associated with the normal theory function (1.31) are WLS, GLS, and RLS estimates of $\mathbf{V}$ in (1.29), which we may call $\hat{\mathbf{V}}_{WLS}$, $\hat{\mathbf{V}}_{GLS}$, and $\hat{\mathbf{V}}_{RLS}$, respectively. Then standard errors for these estimates can be obtained by using $\hat{\mathbf{V}}_{WLS}^{-1}$, $\hat{\mathbf{V}}_{GLS}^{-1}$, or $\hat{\mathbf{V}}_{RLS}^{-1}$, based on (1.30), in place of $\mathbf{V}$ in (1.9). If the multivariate normal distribution is only a rough approximation to the distribution of the data, robust standard errors can be computed instead, as given by (1.11). For example, robust standard errors for the GLS estimator can be computed by evaluating (1.11) at $\mathbf{W} = \hat{\mathbf{V}}_{GLS}^{-1}$ and $\mathbf{V} = \hat{\mathbf{V}}_{ADF}$, the unstructured estimator obtained from (1.17). Model fit under possible distributional misspecification can be evaluated by referring the model test statistic to (1.12) or by using any of the statistics in (1.21)–(1.25), with $\mathbf{W}$ and $\mathbf{V}$ defined similarly. That is, all residual-based statistics, mixture tests, and Satorra–Bentler statistics can be specialized to the normal theory case by defining $\mathbf{W}$ and $\mathbf{V}$ appropriately.

As far as we know, in the context of correlation structure analysis there is no comparative information on the relative performance of the WLS, RLS, and GLS estimators, their estimated standard errors, or the associated test statistics under various conditions, such as small sample size or misspecifications in the model or data distributions. If the distribution of the data is specified correctly, these estimators are asymptotically equivalent, and so are the associated test statistics. Hence, any differences in performance will occur in small samples. In the covariance structure context, we know that GLS (Browne, 1974) tends to overaccept models somewhat at the smallest sample sizes compared to RLS, but we do not know whether this finding will generalize to correlation structures.

1.3.8 Elliptical Theory Estimation and Testing

This approach assumes that the data came from an elliptical distribution, which has its density proportional to

$$|\Sigma|^{-1/2}g\{(\mathbf{x} - \mu)'\Sigma^{-1}(\mathbf{x} - \mu)\} \tag{1.32}$$

The parameters μ and Σ can be interpreted as the location vector and the scale matrix of the vector of variables $\mathbf{x}$, and g is a nonnegative function. As is evident by comparing to (1.26), the elliptical distribution includes multivariate normal as a special case, with $g(t) = e^{.5t}$. Elliptical distributions have heavier or lighter tails than the normal distribution, but remain symmetrically distributed. They are thus a very general class of distributions that can be used to model data with insignificant amount of skewness, but nonzero amount of kurtosis, although they make the assumption that univariate kurtosis is the same for all the variables. Relaxing this assumption results in another estimation method, heterogeneous kurtosis (HK) theory, discussed in the next section.

It can be shown that for elliptical distributions, the general fourth-order moment given in (1.14) can be expressed as (Bentler, 1983)

$$\sigma_{ijkl} = (\kappa + 1)(\sigma_{ij}\sigma_{kl} + \sigma_{ik}\sigma_{jl} + \sigma_{il}\sigma_{jk}) \tag{1.33}$$

where κ is the common kurtosis parameter. When $\kappa = 0$, the distribution has no excess kurtosis, and we obtain (1.27). In heavier tailed distributions (as compared to the multivariate normal), κ is positive, and in lighter tailed distributions, κ is negative. The following relationship also holds:

$$\kappa = \sigma_{iiii}/3\sigma_{ii}^2 - 1 \tag{1.34}$$

for all $i = 1, \dots, p$, where, as before, $\sigma_{iiii} = E(x_i - \mu_i)^4$ and $\sigma_{ii} = E(x_i - \mu_i)^2$. Substituting (1.33) into (1.15), we obtain the standardized fourth-order moment under elliptical theory:

$$\rho_{ijkl} = (\kappa + 1)(\rho_{ij}\rho_{kl} + \rho_{ik}\rho_{jl} + \rho_{il}\rho_{jk}) \ . \tag{1.35}$$

Using (1.35) in (1.16), we obtain an expression for a typical element of $\mathbf{V}$ under elliptical theory as follows:

$$v_{ij,kl(E)} = (\kappa + 1)v_{ij,kl(NT)} \tag{1.36}$$

Writing this compactly, the normal theory and elliptical theory asymptotic covariance matrices of sample correlations are related by $\mathbf{V}_E = (\kappa+1)\mathbf{V}_N$, and only one more parameter is needed to describe possible increase or decrease in variance. Similarly, it follows that $\mathbf{V}^{-1} = (\kappa + 1)^{-1}\mathbf{V}_N^{-1}$. This implies that the elliptical version of the fit function Q_n in (1.7) can be written as

$$Q_{n(E)} = (\kappa + 1)^{-1}Q_{n(NT)} \tag{1.37}$$

which is just a multiple of the normal theory fit function in (1.31). Since κ is a constant, the minimum of (1.37) with respect to θ is identical to the minimum of (1.31), and this will also hold when we replace κ by any consistent estimator $\hat{\kappa}$. This means that any of the weighting methods that we have discussed for normal theory apply directly to elliptical theory, and we may obtain WLS, GLS, RLS, LS, or diagonally weighted functions, estimators, tests, and standard errors. In particular, for any particular choice of estimator, we have (1) the parameter estimator obtained under a particular normal theory method, say $\hat{\theta}_{NT}$ based on RLS, will be identical to the estimators obtained under its counterpart method in elliptical theory, say $\hat{\theta}_E$ under ERLS; (2) the test statistic associated with a particular normal theory method, say RLS, will be related to its elliptical counterpart, say ERLS, by the relation $T_{ERLS} = (\kappa + 1)^{-1}T_{RLS}$;

and (3) the asymptotic covariance matrix of $\hat{\theta}_{NT}$ obtained by any particular normal theory method will be related to the asymptotic covariance matrix of the corresponding elliptical estimator $\hat{\theta}_E$ by $(\dot{\rho}'V_E^{-1}\dot{\rho})^{-1} = (\kappa + 1)(\dot{\rho}'V_{NT}^{-1}\dot{\rho})^{-1}$ in the case of asymptotically efficient estimators and by a similar relation based on specialized versions of (1.11) with inefficient estimators. These relationships are implicit in Mels (2000), and consistent with Shapiro and Browne's (1987) discussion of scale-invariant correlation structures $\Sigma = DP(\theta)D$.

As a result, we can use normal theory correlation methods with elliptical data as long as we have an estimate of the common kurtosis parameter κ. An estimator with optimal properties (Bentler & Berkane, 1986; Berkane & Bentler, 1990) is based on Mardia's (1970, 1974) multivariate kurtosis. Mardia's multivariate kurtosis estimator is given by:

$$m = \frac{1}{N} \sum_{i=1}^{N} \left[(x_i - \bar{x})'S^{-1}(x_i - \bar{x}) \right]^2 - p(p+2) \qquad (1.38)$$

Then a Mardia-based estimate of κ is given by:

$$\hat{\kappa} = m/p(p+2) \qquad (1.39)$$

This estimator was recommended by Bentler and Berkane (1986), Browne (1982, 1984), and Shapiro and Browne (1987). When the data are symmetric but not elliptical, the scaling provided by (1.39) may perform badly (Satorra & Bentler, 1988). Estimators outside the elliptical class (Kano, 1992), such as the multiplier of the Satorra-Bentler scaled test in (1.24), are more robust to violation of ellipticity. That is, using the estimator $\hat{\kappa} = \text{tr}(\hat{U}\hat{V})/d$ in the computation of the elliptical test statistic $T_{ERLS} = (\kappa+1)^{-1}T_{RLS}$ amounts to applying the Satorra–Bentler robust test statistic (1.24) to the normal theory statistic T_{RLS}. Using this same $\hat{\kappa}$ to compute elliptical theory standard errors by correcting the normal theory asymptotic covariance matrix as $(\kappa + 1)(\dot{\rho}'V_N^{-1}\dot{\rho})^{-1}$ may yield improved standard errors, even if these are not the fully general robust standard errors obtained by using (1.11) with $V = \hat{V}_{ADF}$. Of course, the latter requires much heavier computations. These suggestions remain to be studied.

With covariance structure analysis, a disadvantage of the elliptical methods is that they may not be as robust to small violations of distributional assumptions as are normal theory methods. Initially optimistic performance found in simulations (e.g., Harlow, 1985) has become less so (e.g., Boomsma & Hoogland, 2001; Hoogland, 1999). This may be true here also.

1.3.9 Heterogeneous Kurtosis Theory Estimation and Testing

Heterogeneous kurtosis (HK) theory (Kano, Berkane, & Bentler, 1990) generalizes elliptical distribution theory to allow individual variables to have different kurtosis parameters. Normal and elliptical theories can be viewed as special cases of HK, in the sense that if the HK method is used on elliptical or normal data, there is no penalty, and the method specializes as needed. However, HK theory is *not* a generalization of the elliptical theory in the sense that it does not generalize (1.32) to a wider class of densities. Instead, it generalizes the structure of the fourth-order moments implied by elliptical distributions and given in (1.33) while still requiring significantly fewer parameters than (1.16). The class of densities that would generate such a structure, however, is not yet known.

Kano et al. (1990) proposed the following structure for the fourth-order moment:

$$\sigma_{ijkl} = (a_{ij}a_{kl})\sigma_{ij}\sigma_{kl} + (a_{ik}a_{jl})\sigma_{ik}\sigma_{jl} + (a_{il}a_{jk})\sigma_{il}\sigma_{jk} \qquad (1.40)$$

where a_{ij}, elements of a symmetric matrix $\mathbf{A}$, are arbitrary parameters except for the restriction that $\mathbf{V}$ in (1.16) remain positive definite. Substituting (1.40) into the general expression for the standardized fourth-order moment in (1.15), we obtain

$$\rho_{ijkl} = (a_{ij}a_{kl})\rho_{ij}\rho_{kl} + (a_{ik}a_{jl})\rho_{ik}\rho_{jl} + (a_{il}a_{jk})\rho_{il}\rho_{jk} \qquad (1.41)$$

Using (1.41) in (1.16) will give an expression for the typical element of $\mathbf{V}$ structured under HK theory. We omit this expression because it does not yield any additional insights.

This method is called heterogeneous kurtosis because variables' marginal kurtoses, which are allowed to differ, are used to define the a_{ij}'s in (1.41). Define the measure of kurtosis for variable i as $\eta_i^2 = \sigma_{iiii}/3\sigma_{ii}^2$. There are two main possibilities for defining the elements of $\mathbf{A}$. Kano et al. (1990) proposed the arithmetic average method, where

$$a_{ij} = .5(\eta_i + \eta_j) \qquad (1.42)$$

Bentler, Berkane and Kano (1991) proposed the geometric mean method, where

$$a_{ij} = \sqrt{\eta_i\eta_j} \qquad (1.43)$$

Both (1.42) and (1.43) try to approximate the fourth-order moments of the variables with (1.41). The geometric approach holds for a wider variety of nonnormal distributions, and should perform better, although the empirical evidence on this is limited. Hu et al. (1992) found that there was a tendency for HK methodology implemented using (1.42) to produce a chi-square statistic that somewhat overaccepts models. Because the arithmetic mean is always greater than the geometric mean, there should be less overcorrection using (1.43).

As is the case with normal and elliptical theory methods, estimation can be done using weighted, generalized, or iteratively reweighted least squares. In the HKGLS approach, estimation is conducted by minimizing (1.7) with the weight matrix set to $\mathbf{W}_n = \hat{\mathbf{V}}_{HKGLS}^{-1}$, where $\hat{\mathbf{V}}_{HKGLS}$ is obtained by using sample correlations r_{ij} in its HK version. In the HKRLS and HKWLS approaches, model-based estimates are used instead, and $\mathbf{W}_n$ is iteratively updated either as a constant matrix or as one differentiated during optimization. Standard errors can be obtained by using the relevant $\hat{\mathbf{V}}_{HKGLS}^{-1}$, $\hat{\mathbf{V}}_{HKRLS}^{-1}$, or $\hat{\mathbf{V}}_{HKWLS}^{-1}$ in (1.9), and the model fit can be evaluated referring (1.10) to a chi-square distribution with $d = \tilde{p} - q$ degrees of freedom. Robust standard errors and test statistics can also be used. For completeness, we note that we can also define diagonally weighted least squares methods and least squares methods in parallel to those given above. By now it should be clear that the logic of both the correctly specified and the robust methodologies stays the same across methods, and what distinguishes different methods is our belief about the form of in the population, which drives the choice of the weight matrix.

Although the HK methodology was developed over a decade ago and found to be promising in one study beyond that of its developers (Hu, Bentler & Kano, 1992), it seems not to have been studied further. Hence, the limitations of the method are really not well known, even with covariance structures. One abstract problem is that the classes of variable distributions that are covered by HK theory are not really understood. That is, when using HK theory, we cannot start our research reports with, "Assume the data were generated from ... " However, to the extent that the method provides a reasonable approximation to the fourth-order moment structure of the data, it will be preferred to any approach that uses a well-defined multivariate distribution that does not account for the data equally well. The HK methodology has not previously been extended to or tested with correlation structures.

1.3.10   Least Squares Estimation and Testing

The robust methodology developed earlier allows for normal, elliptical, and HK theories to be used even when initial estimation of parameters proceeds differently. In this section we discuss the simplest estimator possible: the one that is obtained by minimizing (1.7) by setting $\mathbf{W}_n = \mathbf{I}$, the identity matrix. This is the least-squares (LS) estimator, familiar from regression, but never optimal in correlation structure analysis. This procedure will simply minimize the sum of squared residuals. The estimator remains consistent, but obviously, using $\mathbf{V}^{-1} = \mathbf{I}$ in (1.9) and referring (1.10) to a chi-square variate will produce incorrect results.

However, we can, instead, compute robust standard errors with (1.11) and all robust test statistics of (1.21)–(1.25). In all these equations it is clear that $\mathbf{W} = \mathbf{I}$. To estimate $\mathbf{V}$ we could use the completely unstructured ADF estimator $\hat{\mathbf{V}}_{ADF}$ given in (1.17), but the resulting statistic is likely not to behave well unless the sample size is very large and the model is simple. But suppose we do believe that our data roughly follow some parametric distribution; for simplicity, let us assume that it is the multivariate normal distribution. Then we can estimate $\mathbf{V}$ as $\hat{\mathbf{V}}_{GLS}$, for instance. This allows us to specialize the residual-based statistic given by (1.22) to normally distributed data as follows:

$$T_{RES(NT)} = n(\mathbf{r} - \rho(\hat{\theta}_{LS}))'\hat{\mathbf{U}}_{V(NT)}(\mathbf{r} - \rho(\hat{\theta}_{LS})) \qquad (1.44)$$

where θ_{LS} is the least-squares estimator obtained by minimizing (1.7) with $\mathbf{W}_n = \mathbf{I}$, and $\hat{\mathbf{U}}_{V(NT)}$ is a consistent estimator of the normal theory residual weight matrix $\mathbf{U}_{V(NT)} = (\mathbf{V}^{-1} - \mathbf{V}^{-1}\dot{\rho}(\dot{\rho}'\mathbf{V}^{-1}\dot{\rho})^{-1}\dot{\rho}'\mathbf{V}^{-1})$, where $\mathbf{V}^{-1}$ is estimated by (1.30) with $\hat{\mathbf{V}}_{GLS}^{-1}$. This statistic has an asymptotic χ_d^2 distribution, as long as the fourth order moments of the data are structured according to (1.29). Some algebra can verify that (1.44) can be computed using only quantities such as functions and gradients available for a given estimation method, say, normal (Bentler 1989, p. 217). Other easy to compute versions of (1.44) are possible, such as those based on elliptical or HK distributions. Standard errors for the least squares estimator are obtained from the relevant specialization of (1.11), that is,

$$\sqrt{n}(\hat{\theta} - \theta) \xrightarrow{d} N(0, (\dot{\rho}'\dot{\rho})^{-1}\dot{\rho}'V\dot{\rho}(\dot{\rho}'\dot{\rho})^{-1}) \tag{1.45}$$

estimating V with any asymptotically efficient estimator under the distribution chosen, such as $V\hat{}_{WLS}$, $V\hat{}_{GLS}$, or $V\hat{}_{RLS}$ in the case of normal theory.

This example demonstrates the versatility of the proposed robust corrections. In principle, any estimator can be used as long as it is consistent and appropriate standard error and test statistics formulas exist. This would include the diagonally weighted estimators that we have introduced above. Of course, the advantage of minimizing (1.7) with $W_n \rightarrow V^{-1}$ is that the resulting estimator is asymptotically efficient (i.e., it has the smallest possible variance of all estimators in its class). That is, if we believe that our data are multivariate normal (or some other distribution), we should use the WLS, GLS, or RLS estimators directly instead of the least-squares robust methodology described here. However, estimators obtained from correctly specified discrepancy functions are only asymptotically efficient. Their behavior in small to medium sample sizes relative to the LS estimator is unknown. The LS estimator, and similarly the diagonally weighted estimators, have the advantage that they are extremely easy to compute. Evaluating the asymptotic loss of efficiency of these estimators—which may be quite small—and the relative performance of the various methods in small samples is still an open problem.

1.4 EXAMPLE

In developing multiple factor analysis, Thurstone (1947) proposed that the method should reproduce correlations from latent factors consistent with acceptably small residual errors. Although Thurstone (1954) has described the idea, Comrey (1962; Comrey & Ahumada, 1965) was the first to implement it and make the method generally available. Comrey's method was sequential: find the minimum residual (minres) factor that best accounts for correlations, residualize, then repeat on the residual matrix for a second factor, and so on. That is, given the correlation matrix R, he found the $p \times 1$ factor loading vector $\hat{\lambda}$ by minimizing the sum of squared residuals of $\{R - \lambda\lambda' - (I - \text{diag}(\lambda\lambda'))\}$, then obtained the residual matrix $R_{res} = (R - \hat{\lambda}\hat{\lambda}')$ and repeated the procedure until k factors were extracted. He later proposed to iterate the k-factor solution (e.g., Comrey & Lee, 1992). In a discussion of the Thurstone (1954) paper, Tukey suggested that it might be better to do the minimization with respect to all the hypothesized factors, but it was not until Boldt (1965) and Harman and Jones (1966) that this was accomplished. In this k-factor minres method, the function (1.7) with $W_n = I$ is minimized for an exploratory

EXAMPLE   **23**

factor model $\mathbf{P} = \mathbf{\Lambda\Lambda'} + \{\mathbf{I} - \mathbf{diag(\Lambda\Lambda')}\}$ with $p \times k$ factor loading matrix $\mathbf{\Lambda}$ [See also Zegers and ten Berge (1983) and Levin (1988)].

The general idea of minimum residual estimation was a break with decades of prior tradition in which the diagonal elements $\mathbf{diag(\Lambda\Lambda')}$ called communalities needed to be estimated or known prior to use of some optimization procedure to obtain a factor solution. However, these authors provided no statistical theory for evaluating the adequacy of the solution and, as noted in the introduction, when Jöreskog (1967) showed how to obtain the covariance structure MLE, the minimum residual approach became neglected theoretically, although it had provided quite nice solutions. Even Harman (1967) was enthused about the fact that the MLE methodology could yield statistical information that was not available with his own minres method. Paradoxically, quite soon, Tucker and Lewis (1973) were proposing that their "reliability" index might provide a rationale for ignoring the statistical conclusions and lead to acceptance of a non-fitting statistical model if it accounted for the correlations well enough. Recently, Jöreskog (2003) showed that a least squares covariance structure factor analysis applied to the correlation matrix obtains the same estimators as the minres method. However, he did not address the correctness of the resulting test statistic and standard errors.

In this chapter we have provided the theory that makes minres acceptable statistically. We illustrate the method with the correlation matrix of eight physical variables shown in Table 1.1. This table is identical to Table 1 of Harman and Jones (1966) except that decimals have been included. Above the diagonal is the correlation matrix to be analyzed. Below the diagonal are the residuals $(r_{ij} = \hat{\rho}_{ij})$ based on a two-factor minres solution. The minres solution minimizes the sum of squares of these residuals. Harman and Jones reported that their function value at the minimum was .01205. Using EQS 6, we analyzed this matrix with two exploratory factors using the normal theory LS method, that is, minimizing $Q_{n(NT)}$ of (1.31) with $\mathbf{P_0} = \mathbf{I}$, and obtained the identical solution as in Table 1.1. The minimum function value was the same as reported by Harman and Jones, and the sum of squares of the factor loadings, also the same, was 5.959. The normal theory test statistic $T_{RES(NT)}$ from (1.44) yielded a value of 77.5, which when referred to χ^2_{13} shows that the null hypothesis that two factors can explain the correlations is not tenable statistically. This agrees with Harman's (1967, p. 229) conclusions based on the ML solution: "For twenty years, two factors had been considered adequate, but statistically two factors do *not* adequately account for the observed correlations based on a random sample of 305 girls." Our RLS statistic, $T_{RLS} = n\hat{Q}_{n(NT)}$, based on (1.31) with $\hat{\mathbf{P}}_0 = \mathbf{diag(P(\theta)^{-1})}$ iteratively updated, agrees, yielding $T_{RLS} = \mathbf{76.5}$, the same conclusion as the least squares analysis. Table 1.1 shows that the residuals from the two factor LS solution are really quite small, and hence even if additional factors are needed statistically, they will not explain very much. Additional discussion of the implications of incompatibility between tests and residuals can be found in Browne et al. (2002). The spirit of minimum residual methodology is, of course, maintained in all correlation structure methods. That is, the diagonal elements of $\mathbf{R}$ and $\mathbf{P}$ play no essential role in the estimation and evaluation of the models.

Table 1.1 Correlations and Residuals for Eight Physical Variables[a]

Variable	1	2	3	4	5	6	7	8
1. Height		.85	.81	.86	.47	.40	.30	.38
2. Arm span	−.01		.88	.83	.38	.33	.28	.42
3. Length of forearm	−.02	.03		.80	.38	.32	.24	.35
4. Length of lower leg	.04	−.02	−.01		.44	.33	.33	.37
5. Weight	.02	−.03	.01	.01		.76	.73	.63
6. Bitrochanteric diameter	.02	−.01	.01	−.03	.01		.58	.58
7. Chest girth	−.02	.00	−.01	.03	.01	−.03		.54
8. Chest width	−.02	.04	−.00	−.02	−.03	.02	.02	

[a] Correlations in upper triangle, residuals in lower triangle.

1.5 SIMULATIONS

To get some idea of how the asymptotic theory for correlation structures described in this chapter actually works in practice, we conducted two small simulations. The first involves a study of the original and structured ADF test statistics; the second involves several of the statistics based on the RLS estimator. We note that these are not comprehensive simulation studies but, rather, illustrations designed to demonstrate some of the methods described. Both simulation studies use only severely nonnormal data, and we focus on the performance of the selected statistics. More extensive simulation work including normal as well as nonnormal data and investigating the performance of parameter estimates and their standard errors will be reported elsewhere.

1.5.1 Data

The same simulated data sets were used for both studies. They were created as follows. First, multivariate normal data were generated from a two-factor model with five indicators per factor, yielding a total of 10 observed variables. Factor loadings were set to .7, correlation between the two factors was set to .4, factors were set to have variances of 1, and all observed variables had population means of 0 and variances of 1. Sample sizes were set to be 250, 400, 1000, and 3000. Five hundred data sets of each size were created. Each data set was then transformed using the nonnormal transformation due to Vale and Maurelli (1983), which transforms variables to specified values of univariate skewness and kurtosis while preserving their covariance structure. We set skewness to 2 and kurtosis to 7 for each variable. Finally, we note that when applied to observed variables as we have done here, the Vale and Maurelli (1983) procedure creates nonnormal data that violate the conditions of asymptotic robustness (e.g., Satorra & Bentler, 1986). This means that normal theory methods will break down and robust statistics are necessary.

1.5.2 Correlation Structure with ADF Estimation and Testing

In the first study, all data sets were analyzed using three versions of the ADF method: (1) using covariance structure analysis [i.e., model was fit to sample covariances, weight matrix was estimated using sample counterparts to (1.14)], (2) using correlation structure analysis with the traditional unstructured ADF estimator [i.e., weight matrix was estimated using (1.17)], and (3) using correlation structure analysis with the structured ADF estimator [i.e., weight matrix was estimating using (1.18), using least squares (LS) estimates in the first stage]. Let us call the three ADF statistics $T_{ADF,COV}$, $T_{ADF,COR}$, and $T_{TADF,COR}$, respectively. We are interested in comparing the performance of the traditional ADF methods with covariances and with correlations, that is, in the relative performance of $T_{ADF,COV}$ and $T_{ADF,COR}$. Based on the work of Mels (2000), the expectation here is that both methods will perform poorly until the sample size is huge, with $T_{ADF,COR}$ being the worst. We are also interested in comparing the two correlation structure methods. Again from the work of Mels (2000), the expectation here is that $T_{TADF,COR}$ will do better than $T_{ADF,COR}$.

Table 1.2 gives acceptance rates of the test statistics when the true model was fit to data, using the nominal .05 cutoff value as a criterion. Under this criterion, the correct model should be accepted about 95% of the time, or in 475 replications. As expected, both $T_{ADF,COV}$ and $T_{ADF,COR}$ performed poorly, with $T_{ADF,COR}$ performing worse. Even at the sample size of 1000, $T_{ADF,COV}$ still has not quite converged to its asymptotic distribution, despite the small number of observed variables, and rejected the correct model about 8% of the time. At a sample size of 3000, its rejection rate finally reached its nominal level. $T_{ADF,COR}$, on the other hand, rejected the correct model 25% of the time at a sample size of 1000, and about 11% of the time at a sample size of 3000. Thus, this statistic in particular may require unrealistically large sample sizes to work well. Its structured counterpart, however, did quite well. $T_{TADF,COR}$ began by overaccepting models at the smallest sample size of 250, only rejecting the correct model 2.6% of the time. At greater sample sizes, however, it rejected at about the right level.

We conclude by not recommending the ADF methodology that uses a completely unstructured estimate of the weight matrix, with either covariance or correlation structures. The structured or two-stage ADF methodology shows promise. However, much more work is needed to verify that our preliminary findings extend to other models, in particular to larger models and to other types of nonnormality. Based on our limited data, we conclude that $T_{TADF,COR}$ performs fairly well but tends to be overly conservative in smaller samples. Following Fouladi's (2000) classification of research situations into accept-support and reject-support, $T_{TADF,COR}$ may be particularly useful when a researcher is trying to reject a model, because then its tendency to overaccept will work against the researcher's working hypothesis.

Table 1.2   Acceptance Rates of Three ADF Test Statistics

	$T_{ADF,COV}$	$T_{ADF,COR}$	$T_{TADF,COR}$
$N = 250$	356	88	487
	71.2%	17.6%	97.4%
$N = 400$	441	212	483
	88.2%	42.4%	96.6%
$N = 1000$	461	375	481
	92.2%	75%	96.2%
$N = 3000$	471	444	473
	94.2%	88.8%	94.6%

1.5.3   Correlation Structure with Robust Least Squares Methods

In the second study, each data set was analyzed with correlation structure analysis using RLS estimation (which is equivalent to ML estimation in covariance structure analysis), followed by computation of the following three robust statistics: the residual-based statistic, T_{RES}, given in (1.22); the residual-based F statistic, $T_{F(RES)}$, given in (1.23); and the Satorra–Bentler scaled chi-square, $\bar{T}_1$, given in (1.24). We considered both the unstructured and structured versions of all three statistics, a total of six statistics. To our knowledge, these statistics have never before been studied in the context of correlation structure analysis. We expect that the unstructured version of T_{RES} will perform similar to the unstructured ADF test statistic, that is, very poorly. We do not have expectations about the remaining five statistics, except to say that they all attempt to improve on the poor anticipated performance of the unstructured T_{RES}.

Table 1.3 gives the number and proportion of acceptance of the correct model out of 500 replications, or out of the number of converged solutions, using the nominal .05 cutoff value as criterion. As expected, the performance of T_{RES} with the traditional ADF estimator is very poor, even at the largest sample size, and it is only a little better than the performance of the unstructured ADF test statistic (column 2 in Table 1.2). Although smaller, the F-statistic is not much of an improvement and does not lead to nominal level acceptance rates at any sample size. This finding differs from the corresponding finding in covariance structure analysis, where the residual-based F-statistic has been shown to perform very well in smaller sample sizes. The scaled chi-square does well for $N = 1000$ and 3000, but not at smaller sample sizes. The second half of Table 1.3 gives the corresponding results for the structured versions of these statistics. As can be seen, the structured T_{RES} now rejects at a nearly optimal rate, so that correcting it further to yield the F-statistic leads to slight overacceptance for intermediate sample sizes. The scaled chi-square also does reasonably well, except at the smallest sample size. Thus, we conclude that using structured estimates of the ADF matrix has the potential to produce better test statistics. When these statistics do deviate from expected nominal levels, it seems to be in the direction of overacceptance. As with the previous study, all these conclusions should be qualified until a more extensive simulation study is conducted, including, in particular, larger models and varying degrees of nonnormality.

Table 1.3   Acceptance Rates of the Robust Test Statistics Based on RLS Estimation

	$\hat{V}_{ADF}$			$\hat{V}_{TADF}$		
	T_{RES}	$T_{F(RES)}$	T_1	T_{RES}	$T_{F(RES)}$	T_1
N = 250	83/474	161/474	417/474	461	474	443
	17.5%	34.0%	88.0%	92.2%	94.8%	88.6%
N = 400	211/494	285/474	446/474	476	489	472
	42.7%	57.7%	90.3%	95.2%	97.8%	94.4%
N = 1000	375	309	478	483	491	485
	75%	61.8%	95.6%	96.6%	98.2%	97%
N = 3000	444	449	472	473	477	473
	88.8%	89.8%	94.4%	94.6%	95.4%	94.6%

1.6 DISCUSSION

Although we have covered a range of methods for estimating and testing correlation structure models, our review does not cover the entire range of methods that have been proposed. For example, in all of the methods we discuss above, Fisher z-transformations of sample correlations r_{ij} can also be used with appropriate transformations of the weight matrix (e.g., Brien, Venables, James, & Mayo, 1984; Neill & Dunn, 1975; Steiger, 1980a,b). The resulting estimators and test statistics can be based on transformed χ^2 theory (Ferguson, 1996). These methods have been evaluated in limited simulation studies (e.g., Fouladi, 2000; Steiger, 1980a,b), and they also appear to be quite promising. However, even z-transformations of correlations imply generally well-behaved sample correlations, but when distributions are far from normal, or contain outliers, it may be preferable to analyze robust estimates of standard Pearson correlations (e.g., Wilcox & Musca, 2001). As in covariance structure analysis, when data are badly distributed, robust estimators of means and covariances (Yuan & Bentler, 2007), and hence correlations, may well provide improved estimation and testing of correlation structure hypotheses. The details of correlation structures based on robust estimates remains to be developed. Finally, there are related problems that we also have not specifically addressed, such as models for the inverse matrix $P^{-1} = P^{-1}(\theta)$, which are relevant to the study of conditional independence models.

In providing our overview of the variety of methods, we have pointed out several places where additional research could be fruitfully directed. We also showed how correlation structure methods could resurrect an old method of factor analysis which seems to have been ignored due to the absence of a statistical theory to accompany it. Finally, we did some simulations of correlation structures under nonnormal distributions. We found that the ADF methodology in correlation structures requires very large sample sizes to work, even larger than those required for covariance structures. We also found that the residual-based statistics and the scaled chi-square that are based on the structured or two-stage estimator perform much better in small sample sizes, although they may have a tendency to be too conservative. Much more work is needed to evaluate other test statistics that were proposed but not studied (e.g., the adjusted chi-square) and to extend the current study to other conditions, such as larger models.

In conclusion, we remind the reader that correlation structure methodology is not necessarily always the most natural methodology for the analysis of multivariate data. For instance, when means and variances are of primary interest, as they might be in growth curve models, mean and covariance structure methods provide a more informative alternative. Nonetheless, in many situations correlation structure models do provide a natural way to specify a variety of hypotheses regarding the structure and interrelations among observed and latent variables. As this chapter has demonstrated, an appropriate statistical theory for this methodology has been developed, continues to grow, and offers new challenges for further research.

Peter M. Bentler
University of California, Los Angeles

Victoria Savalei
University of British Columbia, Vancouver

REFERENCES

Amemiya, Y., & Anderson, T. W. (1990). Asymptotic chi-square tests for a large class of factor analysis models. *Annals of Statistics, 18*, 1453–1463.

Bartholomew, D. (2007). Three faces of factor analysis. In R. Cudeck & R. C. MacCallum (Eds.), *Factor analysis at 100: Historical developments and future directions.* Mahwah, NJ: Lawrence Erlbaum Associates.

Bartlett, M. S. (1950). Tests of significance in factor analysis. *British Journal of Psychology, Statistical Section, 3*, 77–85.

Bentler, P. M. (1983). Some contributions to efficient statistics for structural models: Specification and estimation of moment structures. *Psychometrika, 48*, 493–517.

Bentler, P. M. (1986). Structural modeling and Psychometrika: An historical perspective on growth and achievements. *Psychometrika, 51*, 35–51.

Bentler, P. M. (1989). *EQS structural equations program manual.* Los Angeles: BMDP Statistical Software.

Bentler, P. M. (1990). Comparative fit indexes in structural models. *Psychological Bulletin, 107*, 238–246.

Bentler, P. M. (1994). A testing method for covariance structure analysis. In T. W. Anderson, K. T. Fang, & I. Olkin (Eds.), *Multivariate analysis and its applications*; Vol. 24 (pp. 123–135). Hayward, CA: Institute of Mathematical Statistics.

Bentler, P. M. (2002–2007). *EQS 6 structural equations program manual.* Encino, CA: Multivariate Software (www.mvsoft.com).

Bentler, P. M., & Berkane, M. (1986). The greatest lower bound to the elliptical theory kurtosis parameter. *Biometrika, 73*, 240–241.

Bentler, P. M., Berkane, M., & Kano, Y. (1991). Covariance structure analysis under a simple kurtosis model. In E. M. Keramidas (Ed.), *Computing science and statistics* (pp. 463–465). Fairfax Station, VA: Interface Foundation of North America.

Bentler, P. M., & Dijkstra, T. (1985). Efficient estimation via linearization in structural models. In P. R. Krishnaiah (Ed.), *Multivariate analysis VI* (pp. 9–42). Amsterdam: North-Holland.

Bentler, P. M., & Weeks, D. G. (1980). Linear structural equations with latent variables. *Psychometrika, 45*, 289–308.

Bentler, P. M., & Yuan, K.-H. (1999). Structural equation modeling with small samples: Test statistics. *Multivariate Behavioral Research, 34*, 181–197.

Berkane, M., & Bentler, P. M. (1990). Mardia's coefficient of kurtosis in elliptical populations. *Acta Mathematicae Applicatae Sinica, 6*, 289–294.

Bock, R. D., & Bargmann, R. E. (1966). Analysis of covariance structures. *Psychometrika, 31*, 507–534.

Boldt, R. F. (1965). Factoring to fit off-diagonals. *U.S. Army Personnel Research Office Research Note 161*, p. 22.

Bollen, K. A. (2002). Latent variables in psychology and the social sciences. *Annual Review of Psychology, 53*, 605–634.

Bollen, K. A., & Curran, P. J. (2006). *Latent curve models: A structural equation perspective.* New York: Wiley.

Bollen, K. A., & Stine, R. A. (1993). Bootstrapping goodness-of-fit measures in structural equation models. In K. A. Bollen & J. S. Long (Eds.), *Testing structural equation models* (pp. 111–135). Newbury Park, CA: Sage.

Boomsma, A., & Hoogland, J. J. (2001). The robustness of LISREL modeling revisited. In R. Cudeck, S. du Toit, & D. Sörbom (Eds.), *Structural equation modeling: Present and future* (pp. 139–168). Lincolnwood, IL: Scientific Software International.

Brien, C. J., James, A. T., & Venables, W. N. (1988). An analysis of correlation matrices: Variables cross-classified by two factors. *Biometrika, 75*, 469–476.

Brien, C. J., Venables, W. N., James, A. T., & Mayo, O. (1984). An analysis of correlation matrices: Equal correlations. *Biometrika, 71*, 545–554.

Browne, M. W. (1974). Generalized least squares estimators in the analysis of covariance structures. *South African Statistical Journal, 8*, 1–24.

Browne, M. W. (1977). The analysis of patterned correlation matrices by generalized least squares. *British Journal of Mathematical and Statistical Psychology, 30,* 113–124.

Browne, M. W. (1982). Covariance structures. In D. M. Hawkins (Ed.), *Topics in applied multivariate analysis* (pp. 72–141). London: Cambridge University Press.

Browne, M. W. (1984). Asymptotically distribution-free methods for the analysis of covariance structures. *British Journal of Mathematical and Statistical Psychology, 37,* 62–83.

Browne, M. W. (1987). Robustness of statistic inference in factor analysis and related models. *Biometrika, 74,* 375–384.

Browne, M. W. (1992). Circumplex models for correlation matrices. *Psychometrika, 57,* 469–497.

Browne, M. W., & Cudeck, R. (1993). Alternative ways of assessing model fit. In K. A. Bollen & J. S. Long (Eds.), *Testing structural equation models* (pp. 136–162). Newbury Park, CA: Sage.

Browne, M. W., MacCallum, R. C., Kim, C.-T., Anderson, B., & Glaser, R. (2002). When fit indices and residuals are incompatible. *Psychological Methods, 7,* 403–421.

Browne, M. W., & Shapiro, A. (1986). The asymptotic covariance matrix of sample correlation coefficients under general conditions. *Linear Algebra and Its Applications, 82,* 169–176.

Browne, M.W., & Shapiro, A. (1988). Robustness of normal theory methods in the analysis of linear latent variate models. *British Journal of Mathematical and Statistical Psychology, 41,* 193–208.

Chamberlain, G. (1982). Multivariate regression models for panel data. *Journal of Econometrics, 18,* 5–46.

Cheung, M. W.-L., & Chan, W. (2004). Testing dependent correlation coefficients via structural equation modeling. *Organizational Research Methods, 7,* 206–223

Chou, C.-P., & Bentler, P. M. (1995). Estimates and tests in structural equation modeling. In R. H. Hoyle (Ed.), *Structural equation modeling: Concepts, issues and applications* (pp. 37–55). Thousand Oaks, CA: Sage.

Chou, C.-P., Bentler, P. M., & Satorra, A. (1991). Scaled test statistics and robust standard errors for nonnormal data in covariance structure analysis: A Monte Carlo study. *British Journal of Mathematical and Statistical Psychology, 44,* 347–357.

Comrey, A. L. (1962). The minimum residual method of factor analysis. *Psychological Reports, 11,* 15–18.

Comrey, A. L. & Ahumada, A. (1965). Note and Fortran IV program for minimum residual factor analysis. *Psychological Reports, 17,* 446.

Comrey, A. L., & Lee, H. B. (1992). *A first course in factor analysis*, 2nd ed. Hillsdale, NJ: Lawrence Erlbaum Associates.

Cudeck, R. (1989). Analysis of correlation matrices using covariance structure models. *Psychological Bulletin, 105,* 317–327.

Cudeck, R., & MacCallum, R. C. (Eds.) (2007). *Factor analysis at 100: Historical developments and future directions.* Mahwah NJ: Lawrence Erlbaum Associates.

Curran, P. J., Bollen, K. A., Chen, F., Paxton, P., & Kirby, J. B. (2003). Finite sampling properties of the point estimates and confidence intervals of the RMSEA. *Sociological Methods and Research, 32,* 208–252.

Curran, P. J., Bollen, K. A., Paxton, P., Kirby, J., & Chen, F. (2002). The noncentral chi-square distribution in misspecified structural equation models: Finite sample results from a Monte Carlo simulation. *Multivariate Behavioral Research, 37,* 1–36.

Curran, P. J., West, S. G., & Finch, J. F. (1996). The robustness of test statistics to nonnormality and specification error in confirmatory factor analysis. *Psychological Methods, 1,* 16–29.

Davies, R. B. (1980). The distribution of a linear combination of random variables. *Applied Statistics, 29,* 323–333.

de Leeuw, J. (1983). Models and methods for the analysis of correlation coefficients. *Journal of Econometrics, 22,* 113–137.

Dijkstra, T. (1981). Latent variables in linear stochastic models. Ph.D. dissertation, University of Groningen.

Duncan, T. E., Duncan, S. C., & Stryker, L. A. (2006). *An introduction to latent variable growth curve modeling.* Mahwah NJ: Lawrence Erlbaum Associates.

Enders, C. K. (2002). Applying the Bollen–Stine bootstrap for goodness-of-fit measures to structural equation models with missing data. *Multivariate Behavioral Research, 37,* 359–377.

Farebrother R. W. (1984). The distribution of a linear combination of random variables. *Applied Statistics, 33,* 366–369.

Ferguson, T. S. (1958). A method of generating best asymptotically normal estimates with application to the estimation of bacterial densities. *Annals of Mathematical Statistics, 29,* 1046–1062.

Ferguson, T. S. (1996). *A course in large sample theory.* Boca Raton, FL: Chapman & Hall/CRC.

Fouladi, R. T. (2000). Performance of modified test statistics in covariance and correlation structure analysis under conditions of multivariate nonnormality. *Structural Equation Modeling, 7,* 356–410.

Fouladi, R. T., & Steiger, J. H. (1999). Tests of identity correlation structure. In R. H. Hoyle (Ed.), *Statistical strategies for small sample research* (pp. 167–194). Thousand Oaks, CA: Sage.

Gabler, S., & Wolff, C. (1987). A quick and easy approximation to the distribution of a sum of weighted chi-square variables. *Statistische Hefte, 28,* 317–325.

Guttman, L. (1953). Image theory for the structure of quantitative variates. *Psychometrika, 18,* 277–296.

Guttman, L. (1954). A new approach to factor analysis: The radex. In P. F. Lazarsfeld (Ed.), *Mathematical thinking in the social sciences* (pp. 258–348). Glencoe, IL: Free Press.

Harlow, L. L. (1985). Behavior of some elliptical theory estimators with nonnormal data in a covariance structures framework: A Monte Carlo study. Ph.D. dissertation, University of California, Los Angeles.

Harman, H. H. (1967). *Modern factor analysis*, 2nd ed. Chicago: University of Chicago Press.

Harman, H. H., & Jones, W. H. (1966). Factor analysis by minimizing residuals (minres). *Psychometrika, 31,* 351–368.

Hoogland, J. J. (1999). The robustness of estimation methods for covariance structure analysis. Unpublished Ph.D. dissertation, Rijksuniversiteit Groningen.

Hotelling, H. (1933). Analysis of a complex of statistical variables into principal components. *Journal of Educational Psychology, 24,* 417–441.

Hsu, P. L. (1949). The limiting distribution of functions of sample means and applications to testing hypotheses. *Proceedings of the First Berkeley Symposium in Mathematical Statistics and Probability* (pp. 359–402). Reprinted in *Collected Papers*, New York: Springer-Verlag, 1983.

Hu, L., Bentler, P. M., & Kano, Y. (1992). Can test statistics in covariance structure analysis be trusted? *Psychological Bulletin, 112,* 351–362.

Jennrich, R. I. (1970). An asymptotic chi-square test for the equality of two correlation matrices. *Journal of the American Statistical Association, 65,* 904–912.

Jöreskog, K. G. (1967). Some contributions to maximum likelihood exploratory factor analysis. *Psychometrika, 32,* 443–482.

Jöreskog, K. G. (1969). A general approach to confirmatory maximum likelihood factor analysis. *Psychometrika, 34,* 183–202.

Jöreskog, K. G. (1970). A general method for analysis of covariance structures. *Biometrika, 57,* 239–251.

Jöreskog, K. G. (1971). Simultaneous factor analysis in several populations. *Psychometrika, 36,* 409–426.

Jöreskog, K. G. (1977). Structural equation models in the social sciences: Specification, estimation and testing. In P. R. Krishnaiah (Ed.), *Applications of statistics* (pp. 265–287). Amsterdam: North-Holland.

Jöreskog, K. G. (1978). Structural analysis of covariance and correlation matrices. *Psychometrika, 43,* 443–477.

Jöreskog, K. G. (2003). Factor analysis by MIRES. www.ssicentral.com/lisrel/techdocs /minres.pdf.

Jöreskog, K. G., & Van Thillo, M. (1973). LISREL: A general computer program for estimating a linear structural equation system involving multiple indicators of unmeasured variables. *Research Report* 73–5, Uppsala, Sweden: Uppsala University, Department of Statistics.

Kano, Y. (1992). Robust statistics for test-of-independence and related structural models. *Statistics and Probability Letters, 15,* 21–26.

Kano, Y., Berkane, M., & Bentler, P. M. (1990). Covariance structure analysis with heterogeneous kurtosis parameters. *Biometrika, 77,* 575–585.

Kim, K. H. (2005). The relation among fit indexes, power, and sample size in structural equation modeling. *Structural Equation Modeling, 12,* 368–390.

Krane, W. R., & McDonald, R. P. (1978), Scale invariance and the factor analysis of correlation matrices. *British Journal of Mathematical and Statistical Psychology, 31,* 218–228.

Lee, S.-Y. (1985). Analysis of covariance and correlation structures. *Computational Statistics and Data Analysis, 2,* 279–295.

Lee, S.-Y., & Bentler, P. M. (1980). Some asymptotic properties of constrained generalized least squares estimation in covariance structure models. *South African Statistical Journal, 14,* 121–136.

Lee, S.-Y., & Jennrich, R. I. (1979). A study of algorithms for covariance structure analysis with specific comparisons using factor analysis. *Psychometrika, 44,* 99–113.

Levin, J. (1988). Note on the convergence of minres. *Multivariate Behavioral Research, 23,* 413–417.

Li, L., & Bentler, P. M. (2006). Robust statistical tests for evaluating the hypothesis of close fit of misspecified mean and covariance structural models. *UCLA Statistics Preprint 494,* http://preprints.stat.ucla.edu.

MacCallum, R. C., Browne, M. W., & Sugawara, H. M. (1996). Power analysis and determination of sample size for covariance structure modeling. *Psychological Methods, 1,* 130–149.

Mardia, K. V. (1970). Measures of multivariate skewness and kurtosis with applications. *Biometrika, 57,* 519–530.

Mardia, K. V. (1974). Applications of some measures of multivariate skewness and kurtosis in testing normality and robustness studies. *Sankhya B, 36,* 115–128.

McArdle, J. J., & McDonald, R. P. (1984). Some algebraic properties of the reticular action model for moment structures. *British Journal of Mathematical and Statistical Psychology, 37,* 234–251.

McDonald, R. P. (1975). Testing pattern hypotheses for correlation matrices. *Psychometrika, 40,* 253–255.

McDonald, R. P., & Marsh, H. W. (1990). Choosing a multivariate model: Noncentrality and goodness of fit. *Psychological Bulletin, 107,* 247–255.

Mels, G. (2000). Statistical methods for correlation structures. Unpublished doctoral dissertation, University of Port Elizabeth, Port Elizabeth, South Africa.

Meredith, W., & Tisak, J. (1990). Latent curve analysis. *Psychometrika, 55,* 107–122.

Millsap, R. E., & Kwok, O. M. (2004). Evaluating the impact of partial factorial invariance on selection in two populations. *Psychological Methods, 9,* 93–115.

Mooijaart, A. (1985). A note on computational efficiency in asymptotically distribution-free correlation models. *British Journal of Mathematical and Statistical Psychology, 38,* 112–115.

Neill, J. J., & Dunn, O. J. (1975). Equality of dependent correlation coefficients. *Biometrics, 31,* 531–543.

Neudecker, H., & Wesselman, A. M. (1990). The asymptotic variance matrix of the sample correlation matrix. *Linear Algebra and Its Applications, 127,* 589–599.

Nevitt, J. (2000). Evaluating small sample approaches for model test statistics in structural equation modeling. Unpublished Ph.D. dissertation, University of Maryland.

Nevitt, J., & Hancock, G. R. (2004). Evaluating small sample approaches for model test statistics in structural equation modeling. *Multivariate Behavioral Research, 39,* 439–478.

Olkin, I., & Finn, J. D. (1990). Testing correlated correlations. *Psychological Bulletin, 108,* 330–333.

Olkin, I., & Finn, J. D. (1995). Correlation redux. *Psychological Bulletin, 118,* 155–164.

Olkin, I., & Siotani, M. (1976). Asymptotic distribution of functions of a correlation matrix. In S. Ideka (Ed.), *Essays in probability and statistics* (pp. 235–251). Tokyo: Shinko Tsusho.

Preacher, K. J. (2006). Testing complex correlational hypotheses with structural equation models. *Structural Equation Modeling, 13,* 520–543.

Raykov, T. (2005). Bias-corrected estimation of noncentrality parameters of covariance structure models. *Structural Equation Modeling, 12,* 120–129.

Satorra, A. (2001). Goodness of fit testing of structural equation models with multiple group data and nonnormality. In R. Cudeck, S. du Toit, & D. Sörbom (Eds.), *Structural equation modeling: Present and future* (pp. 231–256). Lincolnwood, IL: Scientific Software International.

Satorra, A. (2002). Asymptotic robustness in multiple group linear-latent variable models. *Econometric Theory, 18,* 297–312.

Satorra, A. (2003). Power of 2 goodness-of-fit tests in structural equation models: The case of non-normal data. In H. Yanai, A. Okada, K. Shigemasu, Y. Kano, & J. J. Meulman (Eds.), *New developments in psychometrics* (pp. 57–68). Tokyo: Springer-Verlag.

Satorra, A., & Bentler, P. M. (1986). Some robustness properties of goodness of fit statistics in covariance structure analysis. *American Statistical Association: Proceedings of the Business and Economic Statistics Section,* pp. 549–554.

Satorra, A., & Bentler, P. M. (1988). Scaling corrections for chi-square statistics in covariance structure analysis. *Proceedings of the American Statistical Association,* pp. 308–313.

Satorra, A., & Bentler, P. M. (1994). Corrections to test statistics and standard errors in covariance structure analysis. In A. von Eye & C. C. Clogg (Eds.), *Latent variables analysis: Applications for developmental research* (pp. 399–419). Thousand Oaks, CA: Sage.

Satorra, A., & Saris, W. E. (1985). Power of the likelihood ratio test in covariance structure analysis. *Psychometrika, 50,* 83–90.

Seber, G. A. F. (2004). *Multivariate observations.* New York: Wiley.

Shapiro, A. (1985). Asymptotic equivalence of minimum discrepancy function estimators to GLS estimators. *South African Statistical Journal, 19,* 73–81.

Shapiro, A., & Browne, M. W. (1987). Analysis of covariance structures under elliptical distributions. *Journal of the American Statistical Association, 82,* 1092–1097.

Shapiro, A., & Browne, M. W. (1990). On the treatment of correlation structures as covariance structures. *Linear Algebra and Its Applications, 127,* 567–587.

Shipley, B. (2000). *Cause and correlation in biology.* Cambridge, UK: Cambridge University Press.

Sörbom, D. (1974). A general method for studying differences in factor means and factor structures between groups. *British Journal of Mathematical and Statistical Psychology, 27,* 229–239.

Spearman, C. (1904). General intelligence objectively determined and measured. *American Journal of Psychology, 15,* 201–293.

Steiger, J. H. (1980a). Testing pattern hypotheses on correlation matrices: Alternative statistics and some empirical results. *Multivariate Behavioral Research, 15,* 335–352.

Steiger, J. H. (1980b). Tests for comparing elements of a correlation matrix. *Psychological Bulletin, 87,* 245–251.

Steiger, J. H. (2005). Comparing correlations: Pattern hypothesis tests between and/or within independent samples. In A. Maydeu-Olivares & J. J. McArdle (Eds.), *Contemporary psychometrics. A festschrift for Roderick P. McDonald.* Mahwah, NJ: Lawrence Erlbaum Associates.

Steiger, J. H., & Hakstian, A. R. (1982). The asymptotic distribution of elements of a correlation matrix: Theory and application. *British Journal of Mathematical and Statistical Psychology, 35,* 208–215.

Steiger, J. H., & Lind, J. M. (1980). Statistically based tests for the number of common factors. Presented at the Psychometric Society, Iowa City, IA.

Swain, A. J. (1975). A class of factor analysis estimation procedures with common asymptotic sampling properties. *Psychometrika, 40,* 315–335.

Tanaka, J. S., & Bentler, P. M. (1985). Quasi-likelihood estimation in asymptotically efficient covariance structure models. *American Statistical Association: 1984 Proceedings of the Social Statistics Section,* 658–662.

Thurstone, L. L. (1935). *The vectors of mind.* Chicago: University of Chicago Press.

Thurstone, L. L. (1947). *Multiple factor analysis.* Chicago: University of Chicago Press.

Thurstone, L. L. (1954). A method of factoring without communalities. In *1954 Invitational conference on testing problems.* Princeton, NJ: Educational Testing Service, pp. 59–62, 64–66.

Tucker, L. R, & Lewis, C. (1973). A reliability coefficient for maximum likelihood factor analysis. *Psychometrika, 38,* 1–10.

Vale, S. G., & Maurelli, V. A. (1983). Simulating multivariate nonnormal distributions. *Psychometrika, 48,* 465–471.

Wilcox, R. R., & Musca, J. (2001). Inferences about correlations when there is heteroscedasticity. *British Journal of Mathematical and Statistical Psychology, 54,* 39–47.

Wiley, D. E. (1973). The identification problem for structural equation models with unmeasured variables. In A. S. Goldberger & O. D. Duncan (Eds.), *Structural equation models in the social sciences* (pp. 69–83). New York: Seminar.

Wiley, D. E., Schmidt, W. H., & Bramble, W. J. (1973). Studies of a class of covariance structure models. *Journal of the American Statistical Association, 68,* 317–323.

Wood, A. (1989). An F-approximation to the distribution of a linear combination of chi-squared random variables. *Communication in Statistics, Part B, Simulation and Computation, 18,* 1439–1456.

Wright, S. (1921). Correlation and causation. *Journal of Agricultural Research, 10,* 557–585.

Wright, S. (1934). The method of path coefficients. *Annals of Mathematical Statistics, 5,* 161–215.

Yuan, K.-H. (2005). Fit indices versus test statistics. *Multivariate Behavioral Research, 40,* 115–148.

Yuan, K.-H., & Bentler, P. M. (1997a). Mean and covariance structure analysis: Theoretical and practical improvements. *Journal of the American Statistical Association, 92,* 767–774.

Yuan, K.-H., & Bentler, P. M. (1997b). Improving parameter tests in covariance structure analysis. *Computational Statistics and Data Analysis, 26,* 177–198.

Yuan, K.-H., & Bentler, P. M. (1998). Normal theory based test statistics in structural equation modelling. *British Journal of Mathematical and Statistical Psychology, 51,* 289–309.

Yuan, K.-H., & Bentler, P. M. (1999). F tests for mean and covariance structure analysis. *Journal of Educational and Behavioral Statistics, 24,* 225–243.

Yuan, K.-H., & Bentler, P. M. (2006a). Structural equation modeling. In C. R. Rao & S. Sinharay (Eds.), *Handbook of statistics, Vol. 26, Psychometrics* (pp. 297–358). Amsterdam: North-Holland.

Yuan, K.-H., & Bentler, P. M. (2006b). Mean comparison: Manifest variable versus latent variable. *Psychometrika, 71,* 139–159.

Yuan, K.-H., & Bentler, P. M. (2007). Robust procedures in structural equation modeling. In S.-Y. Lee (Ed.), *Handbook of latent variable and related models* (pp. 367–397). Amsterdam: North-Holland.

Yuan, K.-H., & Hayashi, K. (2003). Bootstrap approach to inference and power analysis based on three statistics for covariance structure models. *British Journal of Mathematical and Statistical Psychology, 56,* 93–110.

Yuan, K.-H., Hayashi, K., & Bentler, P. M. (2007). Normal theory likelihood ratio statistic for mean and covariance structure analysis under alternative hypotheses. *Journal of Multivariate Analysis, 98,* 1262–1282.

Yuan, K.-H., & Marshall, L. L. (2004). A new measure of misfit for covariance structure models. *Behaviormetrika, 31,* 67–90.

Zegers, F. E., & ten Berge, J. M. F. (1983). A fast and simple computational method of minimum residual factor analysis. *Multivariate Behavioral Research, 18,* 331–340.

CHAPTER 2

OVERVIEW OF STRUCTURAL EQUATION MODELS AND RECENT EXTENSIONS

Contemporary *structural equation models (SEMs)* have their origins in several disciplines. Sewall Wright's (1918, 1921) work on path analysis is perhaps the first work that originated many of the key characteristics that still appear in contemporary SEMs. But current SEMs have evolved, becoming more inclusive and general than even Wright probably ever imagined. SEMs represent a synthesis of knowledge about multivariate analysis from econometrics, psychometrics, sociometrics ("quantitative sociology"), biostatistics, and statistics, although its development over the last 30 years has occurred mostly in the social and behavioral sciences. Indeed, it is only relatively recently that biostatistics and statistics have become interested in SEMs. Blalock (1964) and Duncan (1966) were early influential works that stimulated research in path analysis and SEM-related procedures in sociology and the other social sciences. Two edited books that represent the early-takeoff period of SEMs in the social sciences are Blalock (1971) and Goldberger and Duncan (1973). The LISREL software program (Jöreskog & Sörbom, 1978) was another major turning point that made sophisticated maximum likelihood estimation of SEMs with or without latent variables possible. Since this early period, SEMs have spread through the social and behavioral sciences to marketing, education, public health, and they continue to diffuse into behavioral genetics, epidemiology, biostatistics, social work, and other disciplines. Many more SEM software packages and books have come into being. There is a separate journal (*Structural Equation Modeling*) and a separate listserv (SEMNET) devoted to these models. SEMs represent well-established modeling techniques that continue to evolve and to incorporate a wide variety of multivariate procedures.

SEMs are nearly always multiequation models. The variables in the model are either latent or observed variables where the former are variables for which there are no direct measures (Bollen, 2002). The latent variables are often measured by multiple indicators, and the analysis takes account of measurement error in observed variables so that the relationships of latent variables are estimated while controlling for measurement errors. This ability to consider measurement error and the relation between latent variables simultaneously gives SEMs considerable power and enables more realistic models that do not assume perfect measurement. Considerable confusion surrounds SEMs, what they are, and their capabilities. Part of the confusion is due to the diverse terms used to refer to this model. Covariance structure models, LISREL models, causal models, path analysis, and latent variable models are just a few of the terms. However, SEM has become the most common name for these models. There also are a number of misunderstandings about SEMs that have emerged and contributed to the confusion. For instance, some argue that interactions of variables cannot be included in SEMs; others suggest that the estimators of SEMs are heavily dependent on having all variables from multivariate normal distributions; and still others believe that a model with "latent variables" cannot be taken seriously or that SEMs are a specialized type of model. All of these claims are false even though these ideas are still in circulation among some with little grounding in this area.

To investigate the sources of these misunderstandings would sidetrack us from our primary goals. Here we would like to give an overview of SEMs and to describe some of the recent work that has shown how to incorporate multilevel models, mixture models, item response theory, and sample design into the SEM framework. The latter topics are examples of the expansion of SEMs to areas that until relatively recently were thought to be distinct from or were inadequately considered in SEMs. Our intended audience is quantitative researchers who are not experts in SEMs but who would like to learn something about these models. The chapter also may be of interest to those with some SEM experience, but who are less familiar with these recent extensions.

In the next section we present the general model and notation. This is followed by a brief discussion of the steps involved in structural equation modeling. After this general background, we move to recent extensions of this model in sections on multilevel models, mixture models, item response theory, and sample design. The conclusion follows.

2.1 MODEL SPECIFICATION AND ASSUMPTIONS

We use a modified version of Jöreskog and Sörbom's (1978) LISREL notation for SEMs, [2]the most common notation in the field. In this notation, SEMs are separated into the *latent variable* and the *measurement models*. The latent variable model is

$$\eta_i = \alpha_\eta + B\eta_i + \Gamma\xi_i + \zeta_i \tag{2.1}$$

where η_i is a vector of the latent endogenous variables, α_η a vector of the intercept terms for the equations, B the matrix of coefficients giving the impact of the latent endogenous variables on each other, ξ_i the vector of latent exogenous variables, Γ the coefficient matrix giving the effects of the latent exogenous variables on the latent endogenous variables, and ζ_i the vector of disturbances. The i subscript indexes the case in the sample. We assume that $E(\zeta_i) = 0, \text{COV}(\xi_i', \zeta_i) = 0$, and that $(I - B)$ is invertible.   Exogenous variables are variables that are not explained within the model and that are uncorrelated with all disturbances in the system. Endogenous variables are those that are influenced directly by other variables in the system besides their disturbances. Two covariance matrices are part of the latent variable model: Φ, the covariance matrix of the exogenous latent variables (ξ), and Ψ the covariance matrix of the equation disturbances (ζ). The mean of ξ is μ_ξ.

The latent variable model sometimes is referred to as the structural model. This can give rise to confusion since the parameters from both the latent variable and measurement model are structural parameters, and referring only to the latent variable model as structural can lead to the wrong impression (Bollen, 1989, p. 11). To avoid this problem we refer to the equations that explain the latent endogenous variables as the latent variable model. The *measurement model* links the latent to the observed responses (indicators). It has two matrix equations:

$$y_i = \alpha_y + \Lambda_y\eta_i + \epsilon_i \tag{2.2}$$
$$x_i = \alpha_x + \Lambda_x\xi_i + \delta_i \tag{2.3}$$

where y_i and x_i are vectors of the observed indicators of η_i and ξ_i, respectively; α_y and α_x are intercept vectors; Λ_y and Λ_x are matrices of factor loadings or regression coefficients giving the impact of the latent η_i and ξ_i on y_i and x_i, respectively; and ϵ_i and δ_i are the unique factors of y_i and x_i. We assume that the unique factors (ϵ_i and δ_i) have expected values of zero, covariance matrices of Θ_ϵ and Θ_δ, respectively, and are uncorrelated with each other and with ζ_i and ξ_i.

2.1.1 Illustration of Special Cases

As we mentioned in the introduction, part of the appeal of SEMs is their generality and their ability to incorporate many common statistical models as special cases. In this subsection we illustrate this aspect of SEMs with a few common cases.

---

[2]The intercept notation is a slight modification of the LISREL notation to keep them all as α's.

2.1.1.1 *Simultaneous Equation Models.* Suppose that we assume that there is no measurement error in any of our variables and that there is a single indicator for each latent variable. In this situation, the measurement model becomes

$$\mathbf{y}_i = \mathbf{I}\eta_i \tag{2.4}$$
$$\mathbf{x}_i = \mathbf{I}\xi_i \tag{2.5}$$

In other words, the measured and latent variables are one and the same and we can replace each latent variable with its observed variable counterpart. Doing so leads the latent variable model to become

$$\mathbf{y}_i = \alpha_\eta + \mathbf{B}\mathbf{y}_i + \Gamma\mathbf{x}_i + \zeta_i \tag{2.6}$$

Comparing this to the simultaneous equation model developed in econometrics (e.g., Johnston, 1984), we find that except for some slight differences in notation, the model is identical. Simultaneous equation models are multiequation models that include a system of relationships between the endogenous and exogenous *observed* variables. Recursive and nonrecursive models are two special cases of simultaneous equation models. From the perspective of the general SEM, simultaneous equation models are a special case which assumes that the observed and latent variables are the same and hence that there is no measurement error.

2.1.1.2 *Multiple Regression, ANOVA, and ANCOVA.* We can further specialize the simultaneous equation model in equation (2.6) by assuming that there is a single endogenous variable (i.e., $\mathbf{y}_i$ is a scalar) and hence no endogenous-to-endogenous variable influences ($\mathbf{B} = 0$) and ζ_i is a scalar. This leads to

$$y_i = \alpha_y + \Gamma\mathbf{x}_i + \zeta_i \tag{2.7}$$

a further specialization of the SEM. This equation is equivalent to a multiple regression model. If all the covariates (explanatory variables) in $\mathbf{x}_i$ are dummy variables, we have a model equivalent to ANOVA. If the covariates include dummy variables, continuous variables and their interactions, we have a model equivalent to analysis of covariance (ANCOVA).

2.1.1.3 *Confirmatory Factor Analysis.* The measurement model equations are

$$\mathbf{y}_i = \alpha_y + \Lambda_y\eta_i + \epsilon_i \tag{2.8}$$
$$\mathbf{x}_i = \alpha_x + \Lambda_x\xi_i + \delta_i \tag{2.9}$$

Either of these equations is sufficient to represent any confirmatory factor analysis model. If all variables are deviated from their means such that observed and latent variables have means of zero, the intercept term is not needed and we have, using the $\mathbf{x}_i$ equation,

$$\mathbf{x}_i = \Lambda_x\xi_i + \delta_i \tag{2.10}$$

This is the most common form in which researchers represent the factor analysis model. Given that we need only one of the equations of the SEM, it is easy to see that factor analysis is a special case of the general SEM. What is less obvious, but what will be illustrated in Section 2.2 on multilevel models, is that the CFA model also permits fitting an unconditional growth curve model.

2.1.1.4 *Exogenous* x *Form of Model.*

Another special form of the general model has utility when all of the covariates from the latent variable model are observed. In other words, we have exogenous *observed* covariates that are measured without error (or with negligible error). A frequent situation where this occurs is when we have demographic variables such as age, race, and gender, and the researcher assumes that these contain no measurement error. The measurement model equation for x then becomes $\mathbf{x}_i = \mathbf{I}\boldsymbol{\xi}_i$ and the latent variable model becomes

$$\eta = \alpha_\eta + \mathbf{B}\eta + \boldsymbol{\Gamma}\mathbf{x} + \zeta \qquad (2.11)$$

An interesting aspect of this model is that x can consist of dummy or continuous variables. Estimation occurs *conditionally on* the values of x. This is less restrictive than assumptions that require x and y to have specific distributions, but it does require that there be no measurement error in x. We return to this topic briefly when we discuss estimation.

2.1.1.5 *Other Special Cases.*

The preceding subsections illustrate some common statistical models that are specializations of the general SEM. Other work has shown that time-series models (e.g., Browne & Zhang, 2007), behavioral genetic ACE models (e.g., Neale & Cardon, 1992), and pooled time-series cross-section random and fixed effect models (Allison & Bollen, 1997) are special cases. Later we illustrate recent work that demonstrates additional models that have been incorporated into the SEM framework. Before doing that, we turn to the steps followed for any model that falls under SEM.

2.1.2 Modeling Steps

There are six steps in modeling in SEMs:

1. Model specification

2. Implied moment matrices

3. Identification

4. Estimation

5. Assessing fit

6. Respecification

The following subsections summarize each step.

2.1.2.1 Model Specification.   The first step in structural equation modeling is to formulate the model. A SEM is heavily dependent on the substantive expertise of the modeler in that the modeler must choose the latent and observed variables that belong in the analysis and must specify the relationships of each variable to the others. This step makes it difficult for a statistician, biostatistician, or anyone without training in the substantive area to use SEMs without the guidance of substantive experts. The researcher uses the latent variable model (2.1) and the measurement model, (2.2) and (2.3), to put these substantive ideas into a statistical model suitable for testing. The SEM equations are general. Analysts need to describe the exact composition of the matrices and all parameters that correspond to a particular model. Illustrations of model specifications are provided in the sections that introduce new extensions of the SEM.

2.1.2.2 Implied Moment Matrices.   The covariance matrix and mean vector of the observed variables are the moments of most concern to SEMs. Once a researcher specifies a model structure, this model structure implies that the population means, variances, and covariances of the observed variables are a function of the model parameters. In SEMs, the population mean vector (μ) and the population covariance matrix (Σ) of the observed variables have implied mean vector, $\mu(\theta)$, and an implied covariance matrix, $\Sigma(\theta)$, counterparts where θ is the vector that contains the coefficients, variances, and covariance parameters that are part of the model. The exact nature of $\mu(\theta)$ and $\Sigma(\theta)$ is determined by the model specified by the researcher. There are general definitions of these implied moments that apply to any SEM. The population implied mean vector is

$$\mu(\theta) = \left[\begin{array}{c} \alpha_y + \Lambda_y(I - B)^{-1}(\alpha_\eta + \Gamma\mu_\xi) \\ \alpha_x + \Lambda_x\mu_\xi \end{array} \right] \tag{2.12}$$

The top part of the right-hand-side vector is the implied mean vector for y, and the lower part is the implied mean vector of x. The other symbols were defined in Section 2.1.

Similarly, $\Sigma(\theta)$ is the implied covariance structure, which is a function of the parameters. The implied covariance matrix is fairly complex for the general model, so we partition the matrix to correspond to the implied covariance matrix of y, of x, and of y and x:

$$\Sigma(\theta) = \left[\begin{array}{cc} \Sigma_{yy}(\theta) & \Sigma_{yx}(\theta) \\ \Sigma_{xy}(\theta) & \Sigma_{xx}(\theta) \end{array} \right] \tag{2.13}$$

These parts of the implied covariance matrix are

$$\Sigma_{xx}(\theta) = \Lambda_x\Phi\Lambda_x' + \Theta_\delta \tag{2.14}$$

$$\Sigma_{xy}(\theta) = \Lambda_x\Phi\Gamma'(I - B)^{-1'}\Lambda_y' \tag{2.15}$$

$$\Sigma_{yy}(\theta) = \Lambda_y(I - B)^{-1}(\Gamma\Phi\Gamma' + \Psi)(I - B)^{-1'}\Lambda_y' + \Theta_\epsilon \tag{2.16}$$

These equations show the function of the model parameters that each variance and covariance of the observed variables equals according to the model specification. The equations are sufficiently general that we can find the implied covariance matrix for any SEM. We just substitute the specific matrices for a particular model into these expressions.

These implied moment matrices imply two hypotheses that should be true if the model is valid:

$$\mu = \mu(\theta) \tag{2.17}$$

$$\Sigma = \Sigma(\theta) \tag{2.18}$$

This means that if the model holds, we should be able to exactly reproduce the means, variances, and covariances using the population parameter values of the model. This means that there should be an exact match between the population means, variances, and covariances and those implied by the parameters in the model. An alternative interpretation represents these hypotheses as (Bollen, 1989, p. 263)

$$\mu - \mu(\theta) = 0 \tag{2.19}$$

$$\Sigma - \Sigma(\theta) = 0 \tag{2.20}$$

In other words, the observed variables' mean vector and covariance matrix are predicted exactly by the implied means and implied covariance matrix, leaving zero residuals in the population when the model is valid. The implied moments are helpful in several aspects of SEMs. For one thing, they underlie the estimation process. For another, they help in assessing the fit of a model and possible respecifications. They are also important in aiding the researcher in assessing the identification of a model, the next step in modeling.

2.1.2.3 *Identification.*
For most distributions of observed variables, the means (μ) and covariance matrix (Σ) of the observed variables exist and are at least theoretically available had we the population of cases. In the population, μ and Σ typically have unique values. The existence of these unique values means that μ and Σ are identified. A key question in SEMs is whether the parameters in θ are uniquely determinable using only information from μ and Σ. If so, the model is identified. If not, the model is underidentified. More specifically, if we can solve uniquely for all parameters in θ, writing each element in the vector as a unique function of the elements of μ and Σ, then θ is identified. If it is not possible for one or more parameters to have a unique solution, those parameters and the model in which they appear are underidentified. In small models or when focusing only on parts of a model, it is sometimes possible to use algebraic derivations to solve the equations of $\mu = \mu(\theta)$ and $\Sigma = \Sigma(\theta)$ for elements of θ so that these elements are unique functions of the means, variances, or covariances of the observed variables. However, in large, complex models these algebraic manipulations become unwieldy. Fortunately, there are rules of identification which easier to apply that enable us to detect when parameters of θ have no unique solution. These rules exist for simultaneous equation models (e.g., Wooldridge, 2002), confirmatory factor analysis (e.g., Davis, 1993), and the general SEM model (e.g., Bollen, 1989, pp. 328–333). Although these rules are use-

ful, no one of them covers all possible SEMs. As a result, researchers often turn to *empirical* tests of identification. These are tests of *local* (not *global*) identification and are available in most SEM statistical software packages. Their main drawback is that they assess local identification so that it is possible for a model to be identified locally, but not globally. In practice, the empirical tests work most of the time, but not *all* of the time in detecting underidentified models. The parameters in an identified model are ready to estimate.

2.1.2.4   Model Estimation.

Using sample information, we can estimate the values of θ. The two main classes of estimators are the full-information and the limited-information estimators. Full-information estimators are by far the most typically employed, so we look at these first.

The full-information maximum likelihood (ML) estimator is the default estimator in most SEM software. The fitting function that implements it is

$$F_{ML} = \ln|\Sigma(\theta)| - \ln|S| + \mathrm{tr}[\Sigma^{-1}(\theta)S] - P_z + (\bar{z} - \mu(\theta))'\Sigma^{-1}(\theta)(\bar{z} - \mu(\theta)) \tag{2.21}$$

where S is the sample covariance matrix, $\bar{z}$ is the vector of the sample means of the observed variables ($\bar{y}$ and $\bar{x}$ stacked in a vector), P_z is the number of observed variables, "ln" is the natural log, $|\cdot|$ is the determinant, and "tr" is the trace of a matrix. The ML estimator, $\widehat{\theta}$, is chosen so as to minimize F_{ML}. This is a full-information estimator in that all parameters from all equations are estimated simultaneously. Furthermore, the value of each parameter has an impact on the fitting function, so that information from the system as a whole enters the estimation.

As a ML estimator, $\widehat{\theta}$ has several desirable properties. It is consistent, asymptotic unbiased, asymptotic efficient, has asymptotically normal distribution, and its asymptotic covariance matrix is the inverse of the expected information matrix. The oldest justification for the ML estimator was that all observed variables come from a multinormal distribution. However, Browne (1984) showed that if the variables have no excess multivariate kurtosis, the ML properties continued to hold even if there were multivariate skewness. Other studies have revealed conditions under which the ML estimator remains robust even when there is excess multivariate kurtosis (see, e.g., Satorra, 1990). In addition, if a model has exogenous x variables as in equation (2.11), the distributional assumption for proper asymptotic significance tests is that the disturbance (ζ) comes from a multivariate normal distribution (Jöreskog, 1973). In the event that none of the preceding conditions holds, a researcher has the option of using asymptotic significance tests that take account of excess kurtosis (e.g., Arminger & Shoenberg, 1989; Satorra & Bentler, 1994) or making use of bootstrapping techniques (e.g., Bollen & Stine, 1990, 1993). The implication is that the distribution of the observed variables does not affect the consistency of the estimator (Browne, 1984) and there are alternatives for calculating significance tests.

On the other hand, structural misspecification of a model is a problem, particularly with full-information estimators, in that an error in one part of the system can propagate its effects through the correctly specified parts of the model. Given the pervasiveness of structural misspecifications of at least some part of a model, this tendency is a disadvantage of full-information estimators.   Limited-information estimators are not immune to this problem, but are less susceptible to it. These estimators have an equation-by-equation orientation. Madansky (1964), Hägglund (1982), and Jöreskog (1983) did work on developing limited-information estimators in factor analysis with uncorrelated uniqueness components. Bollen (1996, 2001) has developed an estimator from the two-stage least squares (2SLS) family that applies to factor analysis and to the full latent variable SEMs, including those with correlated unique factors or errors. Bollen's approach depends on using a scaling indicator for each η_i, such that $y_{1i} = \eta_i + \epsilon_i$ where y_{1i} is the vector of scaling indicators and $\eta_i = y_{1i} - \epsilon_i$. Similarly, $x_{1i} = \xi_i + \delta_i$ leads to $\xi_i = x_{1i} - \delta_i$. When η_i is replaced by $(y_{1i} - \epsilon_i)$ and ξ_i is replaced by $(x_{1i} - \delta_i)$ in the latent variable model and the measurement model, the SEM becomes

$$y_{1i} = \alpha_\eta + By_{1i} + \Gamma x_{1i} + \epsilon_{1i} - B\epsilon_{1i} - \Gamma\delta_{1i} + \zeta_i \qquad (2.22)$$

$$y_{2i} = \alpha_{y_2} + \Lambda_{y2}y_{1i} - \Lambda_{y2}\epsilon_{1i} + \epsilon_{2i} \qquad (2.23)$$

$$x_{2i} = \alpha_{x_2} + \Lambda_{x2}x_{1i} - \Lambda_{x2}\delta_{1i} + \delta_{2i} \qquad (2.24)$$

where y_{2i} and x_{2i} are the nonscaling observed variables. Through substitution the latent variables are removed and replaced by observed variables and composite disturbances. Individual equations in this model appear as regression equations. However, using ordinary least squares (OLS) regression is in general not an option since parts of the composite disturbances are correlated with one or more explanatory variables. This correlation renders OLS an inconsistent estimator. To circumvent this problem, the 2SLS estimator (Bollen, 1996) makes use of instrumental variables (IVs) that are selected for each equation. IVs are observed variables from the model that are uncorrelated with the composite disturbance of an equation and that correlate with the problematic variables on the right-hand side of a given equation. The IVs in this approach are *model–implied instrumental variables* in that according to the model structure these variables should satisfy the conditions of an IV for the equation. Bollen & Bauer (2004) describe an algorithm that researchers can use to find the model-implied IVs among the observed variables for a given model structure. Formulas for the 2SLS estimator are in Bollen (1996, 2001), and common statistical packages such as SAS, Stata, or SPSS have procedures that can be adapted for this estimator for latent variable models. The resulting estimator of the coefficients is consistent, asymptotically unbiased, asymptotically normally distributed, asymptotically efficient among limited-information estimators, and has an asymptotic covariance matrix from which standard errors are available for significance testing.

The ML and the 2SLS described above are for continuous variables. There are other estimators for noncontinuous variables that we describe in Section 2.4.

2.1.2.5 *Model Fit.*   After estimation of the model, our attention turns toward assessing its adequacy in describing the data. In SEM, researchers draw a distinction between overall fit and component fit. Considering overall fit first, a key measure is a chi-square test statistic that is based on a likelihood ratio test that compares the hypothesized model to a hypothetical saturated model that has as many parameters to

estimate as there are variances, covariances, and means of the observed variables. The null hypothesis of this likelihood ratio test is $H_0 : \mu = \mu(\theta)$ & $\Sigma = \Sigma(\theta)$. This corresponds to the implied moment hypotheses described earlier. The test statistic is $T_{ML} = (N - 1)F_{ML}$ and it has degrees of freedom $df = \frac{1}{2}P_z(P_z + 3) - t$, where P_z is the number of observed variables and t is the number of free parameters estimated in the model. A statistically significant chi-square test statistic rejects the null hypothesis, which suggests that the model is not correct. However, high statistical power, typical in large samples, often leads to rejection of the null hypothesis even for trivial departures from the null hypothesis. Because of this a wide variety of alternative overall fit measures are available ranging from Bayesian information criterion (BIC) to a root mean squared error of approximation (RMSEA). [See Bollen and Long (1993) and Hu and Bentler (1999) for more details.]   Controversy surrounds the best measures to use. However, good practice suggests reporting the chi-square test statistic along with its degrees of freedom, p-value, and several other overall fit measures to help in the assessment of model fit. The 2SLS estimator has equation-by-equations tests on the suitability of IVs for an equation (Bollen, 1996).

Good overall model fit is a necessary but not sufficient condition for model adequacy. Some models exhibit very good overall fit but lack fit for pieces of the model. For instance, coefficients might have signs different than expected, parameter estimates might not be statistically significant as predicted, or improper solutions such as negative variances or correlations greater than 1 might be present. Not infrequently, a researcher finds either the overall or component fit to be inadequate. In this case, the researcher often moves toward a change in specification of the model.

2.1.2.6 Respecification.

When the original model is judged to be inadequate, researchers often attempt to change the model structure to improve model fit. The source of the revisions range from carefully considered revisions based on substantive expertise on the plausibility of alternative structures to purely empirical search procedures that seek to make changes that will result in the greatest drop of the chi-square test statistic. Empirically based modifications of the model structure include the examination of covariance residuals, setting nonsignificant parameter estimates to zero (or null hypothesis value), and Lagrangian multiplier tests. The latter, also called modification indices, give the expected decrease in the chi square test statistic when a restriction on a constrained parameter is lifted. Each of these empirical approaches to modification is subject to abuse and to generating misleading results. Hence, the models derived in this fashion must be regarded with caution and, ideally, should be applied to new data to determine whether they replicate.

2.1.2.7 Summary of Steps.

In the preceding subsections, we sketched out the steps involved in modeling with SEMs. Although the SEMs cover a large variety and types of models, these steps are applicable in all cases. An intriguing part of the incorporation of additional models into SEMs is that we can follow these steps and gain new insight into the data that is not available in the usual way that researchers approach these models. For instance, the model fit procedures in SEM can reveal problems or desirable aspects of model fit that are not regularly shown in other implementations.

In the next few sections we present some of the more recent models to be incorporated into SEMs.

2.2 MULTILEVEL SEM

Recent decades have seen an increase in the development and application of multi-level models in the social and behavioral sciences (also known as hierarchical linear models, mixed-effects models, variance components models, and random coefficients models). Indeed, in the social and behavioral sciences, these models have become the standard method for analyzing data collected from two or more levels of sampling (e.g., clustered data or longitudinal data). A drawback of these models, however, is that they are (typically) designed only for observed variables, and hence the results can be biased by errors of measurement. To overcome this limitation, much work has been done to integrate multilevel modeling with SEM. In this section we discuss two general approaches to multilevel SEM. We refer to the first approach as the between-and-within specification and the second approach as the random-effects-as-factors specification. We first present each specification approach and then close this section with a summary and comparison of the two approaches.

2.2.1 The Between-and-Within Specification

The between-and-within approach to multilevel SEM can be traced to Goldstein and McDonald (1988) and McDonald and Goldstein (1989), with important contributions from Muthén (1989, 1994), Lee (1990), Muthén and Satorra (1995), Lee and Poon (1998), Bentler and Liang (2003), and Liang and Bentler (2004), among others. Here we provide a brief summary of this approach to multilevel SEM.

With two-level data, such as individuals nested within groups, the observed score for an individual-level variable y can be decomposed into two parts as follows:

$$y_{gi} = u_g + r_{gi} \qquad (2.25)$$

where i and g index individual and group, respectively, u_g corresponds to the group mean, and r_{gi} corresponds to the individual's deviation from the group mean. Here we shall assume that the observed sample of groups has been randomly selected from a population of groups, and that individuals have been randomly selected within groups. As is typical for these models, we will assume that $u_g \sim N(\mu, \sigma_B^2)$, $r_{gi} \sim N(0, \sigma_W^2)$, and $\text{COV}(u_g, r_{gi}) = 0$. Here, σ_B^2 represents the between-groups variance and σ_W^2 represents the within-groups variance.

Extending to a multivariate model, we may now consider sets of individual-level observations $\mathbf{y}_{gi}$ and $\mathbf{x}_{gi}$. Generalizing equation (2.25), we will then have the following:

$$\begin{bmatrix} \mathbf{y}_{gi} \\ \mathbf{x}_{gi} \end{bmatrix} = \mathbf{u}_g + \mathbf{r}_{gi} \qquad (2.26)$$

where $u_g \sim N(\mu, \Sigma_B)$, and $r_{gi} \sim N(0, \Sigma_W)$. Further, group-level characteristics may also have been measured, designated as z_g. These measures contain no within-groups variability, but they do contain between-groups variability and may thus be correlated with the between-groups component of y_{gi} and x_{gi}. To capture these relations, we augment the between-groups covariance matrix following the notation of Liang and Bentler (2004):

$$\widetilde{\Sigma}_B = \text{COV}\begin{bmatrix} z_g \\ u_g \end{bmatrix} = \begin{bmatrix} \Sigma_{zz} & \Sigma_{zy} \\ \Sigma_{yz} & \Sigma_B \end{bmatrix} \tag{2.27}$$

Like the single-level SEM, the μ, Σ_W, and $\widetilde{\Sigma}_B$ matrices can be parameterized similarly to equations (2.12) and (2.13), with the exception that there is a unique equation for each matrix. Hence, the parameter matrices must be subscripted by B or W to indicate whether $\widetilde{\Sigma}_B$ or Σ_W is the referent. This implies that the parameter values, and even the model structure, may differ at the two levels of the model. The same modeling steps outlined for single-level SEMs are thus relevant for these models as well with the caveat that one must specify both a within-groups and a between-groups model. Additionally, it is sometimes of interest to evaluate whether the model structure at the two levels is similar by constraining the values of some parameters in the between-groups model to equal the corresponding values in the within-groups model.

The value of separating within- and between-group effects, long recognized in multilevel modeling, is that one can simultaneously evaluate contextual and individual effects. These effects would otherwise be confounded, leading to the well-known errors of inference known as Simpson's paradox and the ecological fallacy. These errors are avoided in multilevel models by partitioning the effects of predictors into their between- and within-group components. In multilevel SEM this is reflected in different estimates for Γ_B and Γ_W (or other parameter matrices). Relative to conventional multilevel models, however, the key advantage of multilevel SEM is that biases due to measurement error are avoided by estimating these effects for latent predictors and latent outcomes.

Similarly, the factor loadings and even the factor structure can be specified differently for the within- and between-groups levels of the model. The multilevel confirmatory factor analysis model provides a useful example. We can write this model as

$$x_{gi} = u_g + r_{gi} \tag{2.28}$$

where both the between-groups and within-groups variability in the observed measures are assumed to follow a common factor structure:

$$u_g = \alpha_x + \Lambda_B \xi_{Bg} + \delta_{Bg} \tag{2.29}$$

$$r_{gi} = \Lambda_W \xi_{Wgi} + \delta_{Wgi} \tag{2.30}$$

Equation (2.29) indicates that the group means are related to the group-level factors ξ_{Bg} through the factor loading matrix Λ_B. In contrast, equation (2.30) indicates that within-group differences in the values of the indicators are due largely to the influence of the individual-level factors ξ_{Wgi}, transmitted through the loading matrix Λ_W. Note that these equations imply that

$$\Sigma_B(\theta_B) = \Lambda_B \Phi_B \Lambda'_B + \Theta_B \tag{2.31}$$

$$\Sigma_W(\theta_W) = \Lambda_W \Phi_W \Lambda'_W + \Theta_W \tag{2.32}$$

By substituting equations (2.29) and (2.30) into equation (2.28), we can also write the model as

$$x_{gi} = \alpha_x + \Lambda_B \xi_{Bg} + \delta_{Bg} + \Lambda_W \xi_{Wgi} + \delta_{Wgi} \tag{2.33}$$

which corresponds to the expression given by Muthén (1994) and Muthén and Satorra (1989). If we assume that the factor loading matrices are invariant across the two levels of the model (i.e., $\Lambda_B = \Lambda_W = \Lambda$) and that there are no between-group residuals (i.e., $\delta_{Bg} = 0$), we can simplify this model to be

$$x_{gi} = \alpha_x + \Lambda \left(\xi_{Bg} + \xi_{Wgi} \right) + \delta_{Wgi} \tag{2.34}$$

This simplified model, referred to by Skrondal and Rabe-Hesketh (2004) as a *variance components factor model*, has an appealing interpretation. For this model, ξ_{Bg} and ξ_{Wgi} represent between-groups and within-groups variability in the same factors. The total variances (and covariances) of the latent factors are then partitioned into between- and within-groups components through the estimation of Φ_B and Φ_W. This decomposition permits one to determine how much variability in a latent factor is due to group differences versus individual differences within groups: for instance, by calculating an intraclass correlation for the factor. Finally, the assumption that $\delta_{Bg} = 0$ is motivated by viewing variability other than that due to ξ to be purely measurement error. Such random measurement errors should not produce systematic differences among groups. Alternatively, if one takes a more traditional psychometric stance that variability not explained by ξ includes not just measurement error but also true score variability due to factors specific to each indicator, the assumption that $\delta_{Bg} = 0$ is rather strong. Specifically, it requires that only the means of the common factors ξ differ across groups, not the means of the specific factors.

2.2.2 Random Effects as Factors Specification

The second approach to multilevel modeling in SEM explicitly specifies the random effects in the model to be latent factors. Using factor analysis to fit growth curves to longitudinal data has a long history (see Bollen, 2007). However, Meredith and Tisak (1984, 1990) were the first to show how *confirmatory* factor analysis might fulfill this purpose. This approach was developed initially by Meredith and Tisak (1984, 1990) to fit growth curves to longitudinal data. Later, Rovine and Molenaar (2000, 2001) recognized that this approach to fitting random coefficient growth models could also be used to fit other kinds of multilevel models. Bauer (2003), Curran (2003), and Mehta and Neale (2005) completed this generalization by drawing on new methods of estimation for SEM. In what follows, we present this approach by first discussing the general form of the linear mixed model (of which multilevel models are a subclass) and then showing its relation to SEM.

Fundamental to this approach to multilevel modeling in SEM is the recognition that the moment structure implied by the linear mixed model is remarkably similar to the moment structure implied by a CFA. Specifically, the linear mixed model can be written as

$$\mathbf{y}_g = \mathbf{X}_g \boldsymbol{\beta} + \mathbf{Z}_g \mathbf{u}_g + \mathbf{r}_g \qquad (2.35)$$

where $\mathbf{y}_g$ is a vector of observations on a single variable y for the n_g individuals within group g, $\boldsymbol{\beta}$ a vector of fixed effects corresponding to predictors in the design matrix $\mathbf{X}_g$, $\mathbf{u}_g$ a vector of random effects corresponding to predictors in the design matrix $\mathbf{Z}_g$, and $\mathbf{r}_g$ the residuals for the individual observations in group g. Under the usual assumptions that $\mathbf{u}_g \sim N(\mathbf{0}, \boldsymbol{\Sigma}_\mathbf{u})$, $\mathbf{r}_g \sim N(\mathbf{0}, \boldsymbol{\Sigma}_{\mathbf{r}g})$, and that $\mathbf{u}_g$ and $\mathbf{r}_g$ are uncorrelated, the mean vector and covariance matrix for $\mathbf{y}_g$ are implied to be

$$\boldsymbol{\mu}_g = \mathbf{X}_g \boldsymbol{\beta} \qquad (2.36)$$

$$\boldsymbol{\Sigma}_g = \mathbf{Z}_g \boldsymbol{\Sigma}_\mathbf{u} \mathbf{Z}'_g + \boldsymbol{\Sigma}_{\mathbf{r}g} \qquad (2.37)$$

In comparison, let us consider a simple confirmatory factor model for the indicators $\mathbf{y}_i$. The model equation is

$$\mathbf{y}_i = \boldsymbol{\alpha}_y + \boldsymbol{\Lambda}_y \boldsymbol{\eta}_i + \boldsymbol{\epsilon}_i \qquad (2.38)$$

and the implied moment structure for $\mathbf{y}_i$ is then

$$\boldsymbol{\mu}(\boldsymbol{\theta}) = \boldsymbol{\alpha}_y + \boldsymbol{\Lambda}_y \boldsymbol{\mu}_\eta \qquad (2.39)$$

$$\boldsymbol{\Sigma}(\boldsymbol{\theta}) = \boldsymbol{\Lambda}_y \boldsymbol{\Psi} \boldsymbol{\Lambda}'_y + \boldsymbol{\Theta}_\epsilon \qquad (2.40)$$

where $\boldsymbol{\mu}_\eta$ is the vector of means of $\boldsymbol{\eta}$. The mixed-effects model can then be parameterized via a CFA by the following. First, treat the observed values within groups as distinct indicator variables, so that the vector $\mathbf{y}_i$ is composed of these observations. For example, if the maximum number of observations per group is five then five indicator variables would be defined in the vector $\mathbf{y}_i$ (where i now indexes group) and any groups with fewer than five observations would have the extra indicators coded as missing. Second, impose the constraint $\boldsymbol{\alpha}_y = \mathbf{0}$, as no corresponding term is found in the mixed model equation. Third, define latent variables $\boldsymbol{\eta}_i$ to correspond to the random effects $\mathbf{u}_g$ and fix the values of the factor loading matrix $\boldsymbol{\Lambda}_y$ to equal the values of the observed values of the predictors in the design matrix $\mathbf{Z}_g$. Given this substitution, $\boldsymbol{\Psi}$ becomes the covariance matrix of the random effects, taking the place of $\boldsymbol{\Sigma}_\mathbf{u}$ in the mixed model. The covariance matrix $\boldsymbol{\Theta}_\epsilon$ also replaces $\boldsymbol{\Sigma}_{\mathbf{r}g}$ from the mixed model and can be structured in a variety of ways. Finally, the factor means $\boldsymbol{\mu}_\eta$ replace the fixed effects $\boldsymbol{\beta}$ for any predictors included in $\mathbf{X}_g$ that are also included in $\mathbf{Z}_g$ (i.e., that have both fixed and random effects). Any other fixed effects can be included either by specifying additional latent factors with zero variances but estimated means (where the additional factor loadings are defined by the values from $\mathbf{X}_g$) or by incorporating the predictors as exogenous x variables, as discussed by Rovine and Molenaar (2000, 2001) and Bauer (2003).

Originally, only a few specific types of multilevel models could be fit as SEMs using this formulation. The limitations were, first, that each vector $\mathbf{y}_i$ should be of the same length so that a single mean vector and covariance matrix could serve as sufficient statistics for the ML estimator in equation (2.21) and, second, that the design matrix for the random effects, $\mathbf{Z}_g$, should be identical over groups so that it could be specified via the factor loading matrix $\mathbf{\Lambda}_y$. It is unsurprising, then, that the earliest applications of multilevel models as SEMs were growth models (or latent curve models) in which each individual had been observed on the same assessment schedule. As an example, let us consider a simple linear growth model for four observations on variable y taken at equal intervals. The measurement model for this example would be

$$
\begin{bmatrix} y_{1i} \\ y_{2i} \\ y_{3i} \\ y_{4i} \end{bmatrix} = \begin{bmatrix} 1 & 0 \\ 1 & 1 \\ 1 & 2 \\ 1 & 3 \end{bmatrix} \begin{bmatrix} \eta_{1i} \\ \eta_{2i} \end{bmatrix} + \begin{bmatrix} \epsilon_{1i} \\ \epsilon_{2i} \\ \epsilon_{3i} \\ \epsilon_{4i} \end{bmatrix} \tag{2.41}
$$

Here the factor loading matrix consists of a column of ones and a column of time scores, mimicking the design matrix that might be used in a multilevel growth model. Similarly, the latent factors, η_{1i} and η_{2i}, correspond to the intercepts and slopes of the individual trajectories. In comparison to a mixed-model formulation, the factor means for η_{1i} and η_{2i} will equal the fixed effects for the intercept and time, and their covariance matrix will equal the covariance matrix of the random intercept and time effects. Finally, the covariance matrix of the residuals can be structured similar to a multilevel growth model (e.g., $\mathbf{\Theta}_\epsilon = \sigma^2 \mathbf{I}$ to impose independence and homoscedasticity). Specified correctly, there is thus a one-to-one mapping of the parameters of this linear latent curve model and a multilevel linear growth model (Willett & Sayer, 1994).

More recently, two advances in the estimation of SEMs have allowed for the incorporation of a much wider array of multilevel models. First, the maximum likelihood fitting function (2.21), which had previously been specified in terms of the means and covariance matrix, was rewritten in terms of the individual data vectors (Wothke, 2001):

$$
F_{DML} = \sum_{i=1}^{N} [p_i \log(2\pi) + \log |\mathbf{\Sigma}_i(\boldsymbol{\theta})|
$$
$$
+ (\mathbf{y}_i - \boldsymbol{\mu}_i(\boldsymbol{\theta}))' \mathbf{\Sigma}_i(\boldsymbol{\theta})^{-1} (\mathbf{y}_i - \boldsymbol{\mu}_i(\boldsymbol{\theta}))] \tag{2.42}
$$

The subscripting of $\mathbf{\Sigma}_i(\boldsymbol{\theta})$ and $\boldsymbol{\mu}_i(\boldsymbol{\theta})$ by i was critical to allow for partially missing data in the conventional SEM. Specifically, $\mathbf{\Sigma}_i(\boldsymbol{\theta})$ and $\boldsymbol{\mu}_i(\boldsymbol{\theta})$ could be composed of the rows and columns of $\mathbf{\Sigma}(\boldsymbol{\theta})$ and $\boldsymbol{\mu}(\boldsymbol{\theta})$ that corresponded to the present observations in $\mathbf{y}_i$, and thus could differ across units. For the multilevel SEM, this move to a direct maximum likelihood fitting function allows for vectors of observations that differ in length across groups (i.e., unbalanced data).

By reformulating the likelihood in terms of $\Sigma_i(\theta)$ and $\mu_i(\theta)$, it also became possible to allow the factor loading matrices to differ across units. For multilevel SEM, this meant that the core CFA model could be generalized to

$$y_i = \alpha_y + \Lambda_{yi}\eta_i + \epsilon_i \tag{2.43}$$

with implied moment structure

$$\mu_i(\theta) = \alpha_y + \Lambda_{yi}\mu_\eta \tag{2.44}$$

$$\Sigma_i(\theta) = \Lambda_{yi}\Psi\Lambda'_{yi} + \Theta_\epsilon \tag{2.45}$$

The fixed values in Λ_{yi} could then differ across groups to accommodate heterogeneous design matrices, as first noted by Neale, Boker, Xie, and Maes (1999). This second advance opened the way for almost all multilevel models to be estimated as SEMs.

As an example, suppose that students are sampled from multiple schools and their language proficiency and verbal IQ are measured (as in Snijders & Bosker, 1999). The model holds that verbal IQ predicts language proficiency. Additionally, both overall levels of language proficiency and the relation of IQ to language proficiency are hypothesized to vary across schools (i.e., there is both a random intercept and slope). This model can be formulated as

$$\begin{bmatrix} lang_{1i} \\ lang_{2i} \\ \vdots \\ lang_{ni} \end{bmatrix} = \begin{bmatrix} 1 & IQ_{1i} \\ 1 & IQ_{2i} \\ \vdots & \vdots \\ 1 & IQ_{ni} \end{bmatrix} \begin{bmatrix} \eta_{1i} \\ \eta_{2i} \end{bmatrix} + \begin{bmatrix} \epsilon_{1i} \\ \epsilon_{2i} \\ \vdots \\ \epsilon_{ni} \end{bmatrix} \tag{2.46}$$

where η_{1i} and η_{2i} are the random effects. Notice that this model accommodates both different numbers of students sampled per school (reflected in the different length of the vectors and matrices) as well as the fact that different IQ scores are observed for students in different schools. The variable IQ is referred to by Neale et al. (1999) as a *definition variable*, as it is used to define the values of the factor loadings.

Given these computational advances, there are few multilevel models that cannot also be estimated as SEMs (cross-classified models a rare example). More important, by embedding mulilevel models within SEMs, many more modeling possibilities are available. For instance, a measurement model can be specified for the outcome variable (giving the model a form similar to a higher-order factor model) or for predictors without random slopes, eliminating sources of measurement error. Models including both direct and indirect effects of predictors can also be estimated with relative ease. Model specification, identification, estimation, and fit are identical to the single-level SEM when a homogeneous factor loading matrix can be specified. If the factor loading matrix differs by group according to definition variables, the usual test of overall model fit is often no longer applicable. Bauer (2003) and Curran

(2003) provide additional details on these issues and also discuss other extensions of the model.

2.2.3 Summary and Comparison

The two approaches to estimating multilevel SEMs continue to coexist because each offers distinct advantages in certain circumstances. The between-and-within specification assumes that there are two levels of data and that there are no random slopes in the model. Given those restrictions, it is most useful for applications involving cross-sectional data. It allows for a wide variety of model structures and it facilitates the evaluation of within- versus between-group effects. The random effects as factors specification is the standard method for modeling longitudinal data (i.e., latent curve models) and it is useful for any multilevel model that includes a random slope for an observed predictor. Like the between-and-within specification, it allows for the incorporation of measurement models for the outcomes and certain predictors, but it does not easily allow for different within- and between-group factor structures. Both approaches, however, are superior to incorporating no measurement model at all, the standard practice in multilevel modeling. In some situations, the two multilevel SEM specification approaches can be combined. For instance, given longitudinal measures on individuals within groups, the random effects as factors specification can be used to parameterize a growth model and then this can be combined with the between-and-within specification to account for the clustering of individuals within groups (e.g., Muthén, 1997). In sum, there are few remaining limitations to using structural equation models with multilevel data.

2.3 STRUCTURAL EQUATION MIXTURE MODELS

Latent variable models have traditionally been distinguished by two basic characteristics: whether the observed variables are continuous or discrete and whether the latent variables are continuous or discrete (Bartholomew, 1987; Lazarsfeld & Henry, 1968). Crossing these two characteristics results in four types of models. Traditionally, factor analysis and structural equation modeling required continuous measures and assumed that the latent variables (factors) were continuous. Rasch and item response theory models maintained the assumption of continuous latent variables (traits) but were designed for discrete outcomes. In contrast, latent profile and latent class analysis each assume that the latent variables (profiles or classes) are discrete. Latent profile analysis was developed as the continuous observed variable analog to latent class analysis, which required categorical observed variables. Over the years, these distinctions have blurred considerably. As we discuss in a subsequent section, methods for conducting factor analysis and structural equation modeling with discrete outcomes have existed for many years and are closely related to item response theory models. In this section we discuss an extension for structural equation models (and factor analysis) that permits the estimation of both continuous latent factors and discrete latent classes in the same model. We refer to this extension as a structural equation mixture model (SEMM).

2.3.1 The Model

Before indicating the model form, we must make a few preliminary distinctions. First, to simplify the notation, we reserve the vector x for exogenous observed predictors that are assumed to be measured without error (as in the exogenous x form of SEM discussed earlier). All other observed variables will be placed into the vector y (and hence all factors specified as endogenous). Second, although it is not necessary to assume that y is normally distributed in a standard SEM, we will do so here because this assumption will be crucial for the development of the SEMM. For simplicity, we first consider a model without exogenous predictors and then transition to a model with exogenous predictors.

If y is normally distributed, the probability density function (pdf) for y implied by a standard SEM is

$$\phi(\mathbf{y}_i; \theta) = \frac{1}{(2\pi)^{p/2} |\Sigma(\theta)|^{1/2}} \exp\left\{ -\frac{1}{2} [\mathbf{y}_i - \mu(\theta)]' \Sigma(\theta)^{-1} [\mathbf{y}_i - \mu(\theta)] \right\}$$

(2.47)

where p is the number of variables in y and $\mu(\theta)$ and $\Sigma(\theta)$ are the model-implied mean vector and covariance matrix of the SEM specified. But suppose that the population is actually composed of a finite number of latent classes. Each latent class is characterized by its own multivariate normal distribution, with moments structured according to a SEM. Across classes, the parameters characterizing the SEM may differ, or entirely different model structures may be present. The overall distribution of y is then a mixture of the component distributions of these latent classes. The pdf for the mixture is

$$f(\mathbf{y}_i; \pi, \theta) = \sum_{k=1}^{K} \pi^{(k)} \phi^{(k)}\left(\mathbf{y}_i; \theta^{(k)}\right)$$

(2.48)

where $k = 1, 2, \ldots, K$ indexes the latent class and $\pi^{(k)}$ represents the mixing probability, or proportion of the population that belongs to class k (not to be confused with the mathematical constant $\pi = 3.1416\ldots$ in equation (2.47), $\sum_{k=1}^{K} \pi^{(k)} = 1$, $\pi = [\pi^{(1)}, \pi^{(2)}, \ldots \pi^{(K-1)}]'$ and $\theta = [\theta^{(1)}, \theta^{(2)}, \ldots, \theta^{(K)}]'$. The mean and covariance structure of the normal density function $\phi^{(k)}$ are given by the SEM equations

$$\mu^{(k)}(\theta^{(k)}) = \alpha_y^{(k)} + \Lambda_y^{(k)}(\mathbf{I} - \mathbf{B}^{(k)})^{-1} \mu_\eta^{(k)}$$

(2.49)

$$\Sigma^{(k)}(\theta^{(k)}) = \Lambda_y^{(k)}(\mathbf{I} - \mathbf{B}^{(k)})^{-1} \Psi^{(k)}(\mathbf{I} - \mathbf{B}^{(k)})^{-1'} \Lambda_y^{(k)'} + \Theta_\epsilon^{(k)}$$

(2.50)

Here the (k) superscript is used to indicate that the model parameters and structure may differ over classes. This model was first considered by Blåfield (1980), with important subsequent developments due to Yung (1997), Jedidi, Jagpal, and Desarbo (1997a,b), and Dolan and van der Maas (1998). The primary focus of these papers was on identifying unobserved population heterogeneity in the parameters of a common structural model (i.e., a model with similar form across classes). In practice, it is common to assume that both the model structure and many of the parameter values are identical across classes.

Because the SEMM formulated above stipulates that the distribution of y is mixture of normal distributions, this implies that all of the observed y variables are continuous. Dummy-coded exogenous predictors such as race, religion, or gender, then present some difficulty for the model. Realizing this, Arminger and Stein (1997) and Arminger, Stein, and Wittenberg (1999) formulated a conditional version of the model that would allow for the inclusion of exogenous observed covariates of any type. This conditional SEMM may be written

$$f(y_i; \pi, \theta, x_i) = \sum_{k=1}^{K} \pi^{(k)} \phi^{(k)} \left(y_i; \theta^{(k)}, x_i \right) \tag{2.51}$$

where the mean vector and covariance matrix of the normal density function $\phi^{(k)}$ are given by the equations

$$\mu_i^{(k)}(\theta^{(k)}, x_i) = \alpha_y^{(k)} + \Lambda_y^{(k)}(I - B^{(k)})^{-1} \left(\alpha_\eta^{(k)} + \Gamma^{(k)} x_i \right) \tag{2.52}$$

$$\Sigma^{(k)}(\theta^{(k)}, x_i) = \Lambda_y^{(k)}(I - B^{(k)})^{-1} \Psi^{(k)}(I - B^{(k)})^{-1\prime} \Lambda_y^{(k)\prime} + \Theta_\epsilon^{(k)} \tag{2.53}$$

This model has the dual advantages that no specific distribution is assumed for the x variables, permitting the incorporation of dummy-coded categorical predictors, and that the assumption of within-class normality for y is relaxed to within-class conditional normality.

A final contribution to the SEMM was made by Muthén and Shedden (1999) and Muthén (2001), who incorporated a multinomial regression model for the latent class probabilities to permit the prediction of class membership. This extension of the model replaces the term $\pi^{(k)}$ in equation (2.51) with

$$\pi^{(k)}(x_i) = \frac{\exp \left(\alpha_c^{(k)} + \gamma_c^{(k)\prime} x_i \right)}{\sum_{k=1}^{K} \exp \left(\alpha_c^{(k)} + \gamma_c^{(k)\prime} x_i \right)} \tag{2.54}$$

where the coefficients for the last class (the reference category) are fixed to zero for identification purposes (i.e., $\alpha_c^{(K)} = 0$ and $\gamma_c^{(K)} = 0$). This then allows for the inclusion of predictors of class membership.

2.3.2 Estimation

Maximum likelihood estimation via the EM algorithm is the most common method for fitting SEMMs. This method has an intuitive appeal: in the E step, posterior probabilities of class membership are estimated for each case; in the M step, these probabilities serve as case weights for the class-specific likelihoods. Despite this, several limitations of maximum likelihood bear mentioning. The first limitation is that the number of latent classes K cannot be estimated although it is rarely known in advance. As such, common practice is to estimate SEMMs with different values for K and then compare model fit statistics to determine the model with the optimal number of latent classes. This then leads to a second limitation: the likelihood ratio statistic for two models differing by one latent class is not distributed as a chi-square (setting a mixing proportion to zero represents a constraint to a boundary value, violating one of the regularity conditions of the usual likelihood ratio test). An alternative test distribution derived by Lo, Mendell, and Rubin (2001) for simpler normal mixtures has been applied to SEMMs by Muthén (2003), as has the bootstrapping approach of McLachlan (1987). Despite these developments, however, many practitioners rely on comparisons of information criteria such as the BIC. Finally, a third limitation of ML estimation is that the likelihood surface for finite normal mixture models is known to be irregular, riddled with singularities and local solutions. Although the estimation of SEMMs may benefit from the structure imposed on the component distributions, Hipp and Bauer (2006) found that local solutions remain a significant problem.

Relative to ML estimation, Bayesian methods, such as those explored by Zhu and Lee (2001), may offer some advantages. For instance, using a Bayesian approach, it is possible to estimate the number of latent classes directly. Bayesian estimation of SEMMs is, however, nontrivial, and involves a number of other difficulties that must be addressed (such as label switching, when two classes switch identity during estimation).

2.3.3 Sensitivity to Assumptions

By incorporating both discrete and continuous latent variables into the same model, SEMMs provide an incredibly flexible means to fit multivariate data. This flexibility is what makes it possible to examine population heterogeneity in mean and covariance structures. For instance, Jedidi et al. (1997a,b) provides an example evaluating how latent variables representing product features (i.e., cost, quality, promotion, etc.) differentially affected product satisfaction over latent classes representing response-based (rather than predefined) market segments. Similarly, Muthén (2001) provides two examples of mixtures of latent curve models, which he calls *growth mixture models*. In one example, the latent classes define groups of individuals following different trajectories of change in math achievement. In the other, they represent distinct trajectories of heavy alcohol use through time. In each of these examples, the latent classes are interpreted to represent qualitatively distinct subgroups of the population.

The increased flexibility of the SEMM, however, also brings new sensitivities to model assumptions. First, in contrast to the robustness of the conventional SEM, Bauer and Curran (2003, 2004) demonstrated that distributional assumptions are critical for estimating SEMMs. The SEMM requires that the observed indicators in y be distributed as a mixture of (conditional) normal distributions. In addition to implying that each indicator is continuous, this also implies that their aggregate distribution is non-normal. Simply put, if y was normally distributed, there would be no need for more than one normal component distribution to describe the data and the latent classes would be unnecessary. Indeed, if one fits a properly specified model to data generated from a normal distribution, the extension of the model to include two latent classes often results in a degenerate solution in which one class is estimated to have zero members. Thus, for properly specified SEMMs, nonnormality of the y|x distribution is critical for the identification of the latent classes. Unfortunately, however, nonnormality may arise from sources other than the mixture of subgroups. It may reflect ceilings or floors in the measures, the use of ordinal or count data, or simply a skewed distribution in a homogeneous population. The SEMM cannot readily distinguish between these possibilities and will estimate latent classes any time the y|x distribution is nonnormal. Moreover, because the addition of latent classes will allow the model to better approximate the nonnormal distribution of the observed data, model fit comparisons will typically point to the presence of two or more latent classes. Thus, in any given application, the latent classes may represent distinct population subgroups, as desired, or they may instead reflect simple nonnormality in a homogeneous population.

Other than approximating nonnormality, Bauer ad Curran (2004) noted that another possible function of the latent classes in an SEMM could be to accommodate misspecifications in the covariance structure specified. Considering the case of an SEMM without exogenous predictors, Bauer and Curran (2004) noted that the covariance matrix of the observed data, aggregated over classes, could be computed by the equation

$$\Sigma(\pi, \theta) = \sum_{k=1}^{K} \sum_{l=k+1}^{K} \pi^{(k)} \pi^{(l)} \left[\mu^{(k)}(\theta^{(k)}) - \mu^{(l)}(\theta^{(l)}) \right]$$

$$\times \left[\mu^{(k)}(\theta^{(k)}) - \mu^{(l)}(\theta^{(l)}) \right]' + \sum_{k=1}^{K} \pi^{(k)} \Sigma^{(k)}(\theta^{(k)}) \quad (2.55)$$

Of particular importance is the fact that the observed covariances are partially explained by the mean differences between classes, the first term in this equation. As such, if an SEM is misspecified and fails to adequately reproduce the observed covariances, the addition of latent classes that differ in their model-implied means may result in better fit to the data. This is true even if the observed data are normally distributed and the population is homogeneous.

Bauer and Curran (2004) also investigated the impact of nonlinear relationships on the estimation of SEMMs. Although there are now several methods for modeling nonlinear effects among latent variables, most SEMs continue to include only linear effects. If nonlinear effects exist but are not modeled in the SEM, then extending the model to include latent classes will once again improve the fit of the model to the data even in the absence of a true mixture. In this case, the effect estimated within each class serves as a local linear approximation of the true nonlinear function.

In summary, there are at least three instances in which latent classes can serve functions other than that for which they were intended. They may be estimated in the service of approximating nonnormal but homogeneous distributions, to accommodate misspecifications in the covariance structure of the model, or to recover, in piecewise fashion, unmodeled nonlinear effects. Given the great flexibility of the SEMM, it seems probable that the incorporation of latent classes could accommodate other types of misspecifications as well. Thus, although SEMMs offer many new and interesting modeling possibilities, one must take great care in fitting and interpreting these models.

2.3.4   Direct and Indirect Applications

In a monograph on finite mixture modeling, Titterington, Smith, and Makov (1985) make a useful distinction between direct and indirect applications of mixtures which is equally relevant for SEMMs. In direct applications, the motivation is to resolve a true mixture distribution into its component distributions. Goals of direct applications include identifying the correct number of component distributions, estimation and interpretation of the parameters of the component distributions, and assignment of observations to the correct component. Examples of direct applications include instances in which the observations are known or assumed to come from several groups, but group membership (and perhaps even the number of groups) is unknown. In contrast, indirect applications of finite mixture models use the components distributions to capture otherwise intractable features of the data. In this kind of application, the component distributions are a statistical expedience, and are not thought to have any particular meaning or interpretation. Examples of indirect applications include the use of mixture distributions to approximate irregularly shaped distributions.

We have already discussed some of the potential problems with direct applications. Although a direct interpretation of the latent classes of an SEMM may be desired, it is entirely possible that the classes are actually serving one of several indirect functions, such as the approximation of nonnormality, compensation for misspecification of the covariance structure, or the local approximation of nonlinear effects. Thus the assumption that the latent classes reflect true population subgroups is a strong one. Nevertheless, direct applications have been the primary interest of SEMM developers (see, e.g., Dolan & van der Maas, 1998). By contrast, in indirect applications, no inferences are made about population subgroups, and thus concerns about the interpretation of "spurious" latent classes do not apply.

In a recent paper, Bauer (2005) explored two possible indirect applications of SEMMs. Noting that the latent classes of SEMMs may reflect nonlinearity, Bauer (2005) devised a method whereby the locally linear relationships estimated within classes could be aggregated across classes to produce a semiparametric estimate of the true nonlinear function. Relative to other methods for estimating nonlinear effects in SEMs, the indirect application of the SEMM has two advantages. First, the functional form of the relationship need not be known in advance, and second, the distributions of the latent variables need not be normal. The latter advantage relates to the second indirect application explored by Bauer (2005): namely, the construction of semiparametric density estimates for latent variables. Because normal mixtures can approximate a wide variety of continuous distributions, the component distributions of the SEMM can be aggregated to construct plots of the latent variable distributions.

2.3.5 Summary

Structural equation mixture models represent the blending of traditional SEM with latent class analysis. This integration of continuous and discrete latent variable models has made possible many new types of applications. To date, the developers and users of SEMMs have been interested primarily in direct applications, wherein the estimated classes are thought to represent meaningful population subgroups. In drawing this inference, however, one must be aware that the latent classes of the model can also serve several other roles. These other possible functions of the latent classes are exploited in indirect applications, which use the mixture distribution to accommodate otherwise intractable features of the data but do not involve inferences about population subgroups. In either case, there are a number of issues that arise in the fitting of these models that do not arise in the estimation of standard SEMs. These include selection of the appropriate number of classes and the preponderance of local solutions, topics of continuing methodological research.

2.4 ITEM RESPONSE MODELS

To this point we have considered only SEMs for continuous observed variables and have assumed that the relationships between these variables are linear. However, variables are often measured coarsely, on binary or ordinal scales, requiring some additional elaboration of the SEM model. In this section we describe SEM models for categorical variables and their close relationship to item response theory models.

The CFA model presented in equation (2.10) assumes that a linear relationship exists between the measured variables and the latent factors. In order to estimate the parameters of this model using maximum likelihood (ML), it is also necessary to assume that the measured variables jointly follow a multivariate normal distribution, or at least that there is no excessive multivariate kurtosis (see Section 2.1.2.4 for a discussion of distributional assumptions). There are many occasions in the social sciences where one wishes to perform a CFA with measured variables that are *categorical* in nature. One such example is the factor analysis of items on questionnaires or scales. This kind of categorical CFA (CCFA), called *item factor analysis* in some fields, is perhaps one of the most popular uses for CFA. In order to use the CFA model appropriately when the observed data are categorical, two changes must be made to the typical ML-based CFA paradigm.

2.4.1 Categorical CFA

The first change to the basic CFA model presented in equation (2.10) addresses the fact that the relationship between categorical measured variables and latent factors is not likely to be linear. The basic idea is captured when considering the reasons for moving from an OLS regression to something like logistic, or probit, regression. One way to overcome this difficulty in CFA, labeled the *latent response variable approach* by Muthén (1984), is to suppose that underlying the observed categorical response (x) is a latent response variable (x^*) that is continuous and normally distributed. Although the CFA model in equation (2.10) does not hold for the observed categorical data, that is,

$$x_i \neq \Lambda_x \xi_i + \delta_i \qquad (2.56)$$

it is possible that the model could hold for the latent response variable:

$$x^*_i = \Lambda_x \xi_i + \delta_i \qquad (2.57)$$

The final necessary piece is some mechanism to describe how the x^*'s were categorized into the observed x's. This is accomplished via a threshold model which posits that a given x^* assumes the lowest possible category if it is less than some threshold, τ_1. If x^* is greater than τ_1, but less than τ_2, it assumes the second lowest category, and so on. If there are c categories, then $c - 1$ thresholds are necessary to accomplish the categorization.

Another consequence of having categorical measured variables, as detailed in Bollen (1989, p. 434), is that the population covariance matrix of the observed data, Σ, will not be equal to the population covariance matrix of the latent response variables, Σ^*. As the model outlined in equation (2.57) is focused on the latent response variables, it is this covariance matrix that we wish to estimate. Just as Pearson product moment correlations can be used as a measure of association when both variables in question are continuous, *polychoric* correlations can be used as a measure of association when both variables are categorical. A special case of the polychoric correlation, when the two variables in question are dichotomous, is known as a *tetrachoric* correlation.

Building on earlier work by Chistoffersson (1975) and Muthén (1978), Olsson (1979) presented a ML method for computing polychoric correlations. Olsson notes that while it would be optimal to estimate all correlations and thresholds in one step, this proves to be computationally difficult. Rather than obtaining all estimates simultaneously, Olsson describes a two-step method that is still widely used today. In this two-step method, thresholds are obtained based on the univariate marginal distributions. By noting the proportions of respondents choosing each category, it is possible to use the inverse normal cumulative density function (cdf) to determine the appropriate threshold parameters (τ's). As this is all done relative to the standard normal distribution, τ parameters assume that metric. Once the thresholds have been estimated, Olsson's two-stage method computes each bivariate polychoric correlation separately, treating the thresholds as known values. This greatly reduces the computational burden to estimate polychoric correlations, but comes at some cost. Since the correlation matrix is not estimated at a single step but, rather, is composed piece by piece, there is no guarantee that the resulting matrix will be positive definite. A matrix that is non-positive definite (NPD) *cannot* be a proper correlation matrix. Beyond this, NPD matrices can cause serious problems at later stages in the analysis (Wothke 1993).

Once a researcher has obtained polychoric correlations, the hurdle of the linearity assumption in the original CFA model has been largely overcome. Unfortunately, we still must find some way to obtain *estimates* of the parameters of the model of interest. As mentioned above, the estimator of choice (ML) makes assumptions that are not likely to be satisfied when the data in question are categorical.

2.4.2 CCFA Estimation

While the introduction of Olsson's two-step method for the estimation of polychoric correlations allows one to reestablish the plausibility of a linear relationship (between the latent factors and the latent response variables), it does not have any effect on the joint distribution of the observed data. One way to overcome this difficulty is to move from ML to a different method of estimation that does not make as strict assumptions about the distribution of the measured variables. One popular choice has been *weighted least squares* (WLS), which has also been called *asymptotic distribution free estimation* (ADF; Browne, 1974). WLS makes much less demanding assumptions about the measured variables, requiring only that the "eighth order moments of the observed variables' distribution are finite" (Bollen, 1989, p. 426). The WLS fitting function can be written as

$$F_{WLS} = [s - \sigma(\theta)]'\mathbf{W}^{-1}[s - \sigma(\theta)] \tag{2.58}$$

where s is a vectored version of $\mathbf{S}$, $\sigma(\theta)$ is a vectored version of $\Sigma(\theta)$, and $\mathbf{W}$ is a positive-definite weight matrix of appropriate rank. Most of the discussion in the literature involving WLS deals with nonnormal continuous measured variables or categorical variables. In fact, as described by Browne (1974), WLS is a very general estimator that has as special cases many of the estimators most often used in SEM. If $\mathbf{W}$ is based on the observed covariance matrix, the resulting estimator is generalized least squares (GLS). If the model-implied covariance matrix is used to construct the weight matrix, the ML estimator is obtained. To obtain unweighted least squares (ULS), one simply uses an identity matrix as the weight matrix.

When the measured variables in question are categorical, the appropriate WLS fit function is

$$F_{WLS_C} = [r - \rho(\theta)]'\mathbf{W}^{-1}[r - \rho(\theta)] \tag{2.59}$$

where the observed (s) and implied [$\sigma(\theta)$] vectors of covariances have been replaced by observed (r) and implied [$\rho(\theta)$] vectors of correlations. In this instance the observed correlations will be polychoric correlations. When using polychoric correlations, the appropriate weight matrix is the asymptotic covariance matrix of the polychoric correlations. This matrix contains the variances and covariances of the correlation estimates. The size of the asymptotic covariance matrix can cause problems, even with relatively small numbers of items. It can be difficult to obtain reasonable estimates of the asymptotic covariance matrix in such cases and even more difficult to invert, as is required in equation (2.59). Building on the suggestion of Christoffersson (1975), a number of researchers have suggested using only the diagonal of the asymptotic covariance matrix as the weight matrix in equation (2.59). This strategy, called *diagonally weighted least squares* (DWLS) or *robust weighted least squares*, has proven very effective in practice (Flora & Curran, 2004). As discussed in Muthén, du Toit, and Spisic (in press) using the diagonal of the asymptotic covariance matrix as the weight matrix does create problems for computing standard errors and test statistics, as it is no longer the optimal weight matrix. Corrections, such as the one discussed by Satorra and Bentler (1994), can be applied that also have been shown to perform well (Flora & Curran, 2004).

Although other methods exist for estimating the parameters of a CCFA model in the SEM context, the procedure outlined above is still the most widely used. For an overview of more recent advances in estimation for CCFA model parameters in an SEM framework see Jöreskog and Moustaki (2001), Skrondal and Rabe-Hesketh (2004) and Wirth and Edwards (2007).

2.4.3 Item Response Theory

Item response theory (IRT) is a collection of models that attempt to describe the process by which individuals respond to items. Unlike the development of CCFA described above, IRT was developed specifically to deal with categorical data. Two widely used IRT models, especially in fields outside education, are the *two-parameter logistic model* (2PL) and the *graded response model* (GRM; Samejima, 1969). The 2PL model, which is used with dichotomous item responses, is typically written as

$$P(x_j = 1|\xi) = \frac{1}{1 + \exp[-Da_j(\xi - b_j)]} \tag{2.60}$$

where a_j is a slope parameter, b_j a severity parameter, D a scaling constant, and ξ represents the latent construct being measured. [3] As presented here, this is a unidimensional model, meaning that one and only one latent construct is being measured. The scaling constant puts the parameters of this logistic model into the scale of the normal ogive model. This scaling parameter is something of a historical artifact but is still present in some of the more widely used IRT software packages. Many of the original developments in IRT used the normal ogive model, which is similar to probit regression. The logistic model, originally proposed by Birnbaum (1968), was adopted for its computational convenience and because the results matched those obtained from the more complex normal ogive model when the scaling constant (typically, 1.7) was used. The logistic form has been the dominant form in the IRT realm

---

[3] In the IRT literature θ typically represents the latent variable instead of the ξ that we have shown here. Earlier in this chapter we used θ to represent a model parameter in conformity with the notation typical in SEM. To avoid confusion we use ξ as the latent variable instead of the θ.

since that time. The GRM, which is appropriate for ordered categories, is defined as

$$P(x_j = c|\xi) = \frac{1}{1 + \exp[-a_j(\xi - b_{jc})]} - \frac{1}{1 + \exp[-a_j(\xi - b_{jc+1})]}$$
(2.61)

where all parameters are as defined previously. Unlike the 2PL model, which has only one severity parameter, the GRM has $c - 1$ severity parameters, where c is the number of response categories.

The current standard estimation method for IRT model parameters is maximum marginal likelihood with an EM algorithm (MML/EM), which was developed by Bock and Aitkin (1981) building on earlier work by Bock and Lieberman (1970). This method of estimation is often referred to as a full-information estimator as it analyzes raw data rather than summary measures such as covariance or correlation matrices. Although this estimation method has its advantages, there are associated costs which will be discussed in a subsequent section.

The astute reader will no doubt have noticed some similarities to the description of IRT just given and the discussion of SEM-based CCFA that preceded it. We now turn our attention to a more detailed discussion of the relationship between IRT and SEM-based CCFA.

2.4.4 CCFA and IRT

Relationships between IRT and one-factor CCFA models were noted early on (Lord & Novick, 1968) and proven formally for several cases (Takane & de Leeuw, 1987) almost two decades ago. The work by Takane and de Leeuw (1987) demonstrated the analytic links between the historical normal ogive version of the 2-parameter model and GRM with a one-factor SEM-based CFA using polychoric correlations. The fact that the polychoric correlations assume that the underlying response variables are normally distributed is one of the reasons that Takane and de Leeuw focused on the normal ogive, rather than logistic, forms of the two-parameter and graded response models. As noted by the authors of that work, the empirical results of the SEM-based CCFA should match those of the logistic IRT approach, to the extent that the scaling constant is successful at putting the IRT parameters back onto the normal metric.

To convert factor loadings (λ's) and thresholds (τ's) to slopes (a's) and severity parameters (b's), one can use

$$a_j = \frac{\lambda_j}{\sqrt{1 - \lambda_j^2}} D \quad \text{and} \quad b_j = \frac{\tau_j}{\lambda_j}$$
(2.62)

and to convert IRT-based parameters to SEM-based parameters use

$$\lambda_j = \frac{a_j/D}{\sqrt{1 + (a_j/D)^2}} \quad \text{and} \quad \tau_j = \frac{(a_j/D)b_j}{\sqrt{1 + (a_j/D)^2}}$$
(2.63)

If the IRT parameters in question come from software that uses the scaling factor, it can be omitted from the equations above. Also, note that if there are more than two categories, each appearance of b_j and τ_j should receive an additional subscripted c.

The work of Takane and de Leeuw, along with the conversion equations they provided, make a strong case for the deeply routed methodological similarity of the SEM- and IRT-based approaches to CCFA. For the 2PL and GRM, as noted by Takane and de Leeuw, any differences in parameter estimates comes from the estimation method (and to a very small extent the scaling constant). However, as we will soon see, the impact of the method of estimation can be nontrivial.

2.4.5 Advantages and Disadvantages

Despite the overlap between the SEM and IRT approaches to CCFA, there are still advantages and disadvantages to each that may make one preferable to the other, depending on the context. At the time of this writing, the SEM-based approach is clearly advantageous when dealing with multiple latent factors. Although in some instances, as when there is independent clustering, an IRT-based approach can still be used by treating each factor separately. However, this ignores the inter-factor correlations, which deprives the user of information and omits information that could be used in the estimation process. Although software exists to perform multidimensional IRT analyses using MML/EM (TESTFACT; Bock et al., 2002), it is focused primarily on exploratory analyses and has no provisions for items with more than two categories. In addition to difficulties dealing with multiple factors, IRT models have historically not had usable measures of fit. Although recent work by Cai, Maydeu-Olivares, Coffman, and Thissen (2006) is extremely promising in this regard, SEM-based CCFA has a wide array of fit indices to chose from.

Despite these two significant advantages to using an SEM-based item response model, there are some drawbacks as well. The 2PL and GRM are just two of the many IRT models that exist, yet these two, along with the simpler 1PL model, are the only item response models available using the SEM approach. Although these are very popular IRT models, there are many others that are widely used (e.g., the 3PL model) or that are becoming more popular (e.g., unfolding models). Additionally, the reliance on polychoric correlations necessary to implement the SEM-based item response models can cause serious difficulties in practice. Methods to deal with missing data in this case are not well developed. Many users are still forced to use listwise deletion to obtain polychoric correlations and the asymptotic covariance matrix (or its diagonal). Also, there is no guarantee that the resulting polychoric correlation matrix will be a proper correlation matrix, which renders it of dubious utility for further analyses.

With few exceptions, many of the advantages or disadvantages for each approach to CCFA are matters of implementation, not definitive features of the modeling framework. Although it is always difficult to predict the future of any methodology, the current trends in psychometric research and software development suggest that the already porous boundaries between SEM-based CCFA and IRT-based CCFA will be further eroded (Wirth & Edwards, 2007). For the SEM-based models, advances in estimation continue to span the gap between the current polychoric/WLS strategy and full-information ML procedures. From the side of the traditional IRT framework, recent experiments with Markov chain Monte Carlo (MCMC) estimation have improved that framework's ability to address previous weaknesses, such as multiple factors (Edwards, 2005) and model fit (Sinharay, 2005). It seems possible that a decade from now the distinction currently made between SEM the and IRT approaches to CCFA will have vanished completely.

2.5 COMPLEX SAMPLES AND SAMPLING WEIGHTS

Many samples in the social and behavioral sciences and almost all large-scale survey data contain such features as nesting and unequal probabilities of selection. These types of samples pose problems for standard statistical analysis, including SEMs, because these methods assume that observations are selected independently and with equal probability. In this section we discuss the specific estimation problems for SEMs and some of the modeling techniques and estimation corrections that can be employed for proper inference when analyzing samples with nesting and unequal probabilities of selection.

2.5.1 Complex Samples and Their Features

Simple random sampling (SRS) designs whereby observations are selected independently with equal probability are assumed implicitly when analyzing SEMs. However, in practice, data sets that are selected for inference to national populations and other large groups are generally not selected using SRS. Samples that are not SRS termed *complex samples* typically involve clustering and/or stratification of observations. Complex samples may also contain observations selected with unequal probability. The development of alternative probability-based random sampling methodologies was motivated by the lack of extant population sampling frames as well as other practical constraints to sampling, such as financial and time limitations. The probability sampling methodologies that have developed typically include one or more of the following components: stratification, multiple stages of selection, cluster sampling, and unequal probabilities of selection. Each of these features is described subsequently. Although some sampling designs are more common than others, there are myriad variations of design. Cochran (1977), Kish (1965), Levy and Lemeshow (1999), and Lohr (1999) provide descriptions of the most commonly used sampling designs and their components.

2.5.1.1 Stratification. Stratification involves separate random selections from partitions, or strata, of the population. For example, the population of adults living in the United States is separated or stratified into those living in each region of the country, and a separate sample of individuals from each of the regions is selected and subsequently pooled into one large sample representing the entire nation. Stratification is typically used to ensure coverage of the entire population, to control sample sizes for subclasses, and to optimize efficiency of estimation. Improved efficiency may occur when observations within strata are more homogeneous than the overall population or when larger samples are drawn from strata with more heterogeneous elements and smaller samples are drawn from strata with more homogeneous elements. Stratification does not affect SEM parameter estimates but may affect the standard errors of the parameter estimates when utilized in variance estimation. However, stratification may be associated with unequal selection across strata, which does affect parameter estimates.

2.5.1.2 *Multiple Stages of Selection.* Multiple-stage selection involves selecting sampling elements in stages where at the final stage the observations of interest are selected. For example, selection of a sample that represents the population of adults living in the United States might involve selection of census block groups first, households within block groups second, and selection of adults within households last. The first-stage sampling elements are referred to as the *primary sampling units* or PSUs, the second-stage elements are referred to as the *secondary sampling units* (SSUs), and so on. Multiple-stage selection does not affect SEM estimation directly except when it results in clustering or unequal selection probabilities at one or more stages.

2.5.1.3 *Clustering.* Clustering occurs when multiple sample elements are selected from the sample elements of a previous stage, where the higher-stage sample elements are the clusters. Continuing with the example from above, census block groups are clusters if they include more than one household sample element, and households are clusters if they include more than one adult observation in the sample. Clustering results in the correlation or nonindependence of sample elements within cluster. The degree of correlation (or clustering) differs by the characteristics of the observations of analysis. Correlations of observations within a cluster do not affect SEM parameter estimates, but tend to increase the standard errors of the parameter estimates and test statistics.

2.5.1.4 *Unequal Probabilities of Selection.* A sample feature that may result from a complex sampling process is the unequal probability of inclusion of observations into the sample. Specifically, the probability that observations appear in the sample is not equal to the probability that they appear in the population. Unequal selection probabilities may occur purposely: for example, if households are selected at a higher rate from within poorer census block group strata. Or, unequal probability of selection may be a result of other design elements, such as multiple stages of selection: for example, choosing a single adult from households with various numbers of adults. Unequal inclusion in an achieved sample may also be caused by nonresponse and attrition. Unequal selection probabilities may be the most influential sample feature for SEMs because it potentially results in bias for parameter estimates, standard errors, and test statistics. These biases may be prevented with the use of probability weights in SEM estimation.

2.5.2 Probability (Sampling) Weights.

Probability weights are most simply the inverse of the probability of selection of observations into the sample. However, probability weights may include more than just the inverse of the known selection probability resulting from the selection process. They also often contain poststratification and nonresponse weighting corrections. These adjustments improve weights and, consequently, the quality of estimates because they correct for other deficiencies in the sample, such as unit nonresponse.

A sampling weight is inversely proportional to the selection probability of a sample observation. Thus, the unadjusted or "raw" weight for observation i is

$$\omega_i = \frac{1}{p_i} \tag{2.64}$$

where p_i is the selection probability for observation i. The goal is to weight sample elements so that the sample distribution accurately represents the population distribution. Weights estimate the number of population elements, N, that are represented by each sample observation. So the weight for observation i estimates the number of population elements represented by observation i and the sum of the weights is an estimate of the total population size, $\sum_{i=1}^{n} \omega_i = \hat{N}$.

Probabilities of selection and weights are typically specific to strata in a stratified sample and clusters in a cluster sample, or both. Stratification weights are calculated for sample strata where sampling units have stratum-specific probabilities of selection. In cluster sampling the probability of selection for a sample element is equal to the probability of selection for the cluster. In multistage cluster selection the probability of selection is a multiplicative factor of the probability of selection at each stage of selection when each stage is selected independently.

One example is a two-stage cluster sample with SRS at each stage. For this design, the probability of selection for a sample element within a cluster is

P (ith element in the jth cluster is selected)

$= P$ (jth cluster selected) P (ith element selected $|j$th cluster selected)

$$= \frac{m}{M} \cdot \frac{n_j}{N_j} = \frac{mn_j}{MN_j} \tag{2.65}$$

where m is the number of clusters in the sample, M the number of clusters in the population, n_j the number of sample elements in cluster j, and N_j the number of population elements in cluster j. In this case the weight is

$$\omega_{ij} = \frac{MN_j}{mn_j} = \frac{1}{p_j p_{ij}} \tag{2.66}$$

where p_{ij} is the conditional probability of individual i selected in cluster j and p_j is the probability that cluster j is selected. Often, stratification is combined with cluster sampling and multiple stages of selection in a sampling procedure, further complicating the composition of the weights.

2.5.3 Violations of SEM Assumptions

2.5.3.1 Clustering and Stratification.
Clustering violates assumptions of independently selected observations because observations within clusters are correlated with one another. Therefore $COV(\zeta_{gi}, \zeta_{gj}) \neq 0$, $COV(\epsilon_{gi}, \epsilon_{gj}) \neq 0$, $COV(\delta_{gi}, \delta_{gj}) \neq 0$, where g represents clusters and i and j represent observations, with $i \neq j$. The result of this violation on estimation is that the standard errors of parameter estimates are underestimated due to the underestimated population variances, $VAR(\zeta_i)$, $VAR(\epsilon_i)$, and $VAR(\delta_i)$. The degree to which this violation biases standard error estimates is directly related to the degree of clustering, which is a function of the ratio of the between-cluster variation, σ_B^2, to the within-cluster variation, σ_W^2, of an endogenous variable. We can get an idea about the degree of clustering by considering the *intraclass correlation coefficient* (ICC):*Intraclass correlation coefficient*

$$\text{ICC} = \frac{\sigma_B^2}{\sigma_B^2 + \sigma_W^2} \tag{2.67}$$

For both clustering and stratification, it is typically although not necessarily true, that disturbances in both the measurement and the latent variable models violate assumptions of constant error variance; that is, for each disturbance $E\left(\zeta_i^2\right) \neq VAR\left(\zeta_i\right)$, $E\left(\epsilon_i^2\right) \neq VAR\left(\epsilon_i\right)$, and $E\left(\delta_i^2\right) \neq VAR\left(\delta_i\right)$. This happens when the variation of observations within clusters or strata differs across clusters or strata: for example, $E\left(\zeta_{gi}^2\right) = VAR\left(\zeta_{gi}\right)$, but $VAR\left(\zeta_{gi}\right) \neq VAR\left(\zeta_{ki}\right)$ where g and k are clusters and $g \neq k$. This violation will bias standard error estimates and test statistics. The direction of bias depends on the pattern of heteroscedasticity.

2.5.3.2 Unequal Probabilities of Selection.
The usual results from SEM assumes that observations are randomly selected with equal probabilities. It is well known that selection of observations from a population with unequal probabilities of selection may result in sampling distributions that do not represent the population distribution. This would produce sample estimates such as means, covariances, and variances that are biased and inconsistent. SEM relies on S and $\bar{z}$ as unbiased, consistent estimators of the population covariance matrix Σ and mean vector μ, respectively. The quality of estimation, $\Sigma(\hat{\theta})$ and $\mu(\hat{\theta})$, of the population model, $\Sigma(\theta)$ and $\mu(\theta)$, relies on the quality of S and $\bar{z}$ as estimators. If S is biased, SEM model parameters are also biased. Parameter estimates will be affected by unequal probabilities of selection if the distribution of endogenous variables are biased after conditioning on the other variables in the model.

In addition, the assumption that $E(\zeta_i) = E(\epsilon_i) = E(\delta_i) = 0$ is potentially violated if observations from certain groups are oversampled and then a marginal model that pools over the groups is fit to the data. This results in too many or too few error terms for one segment of the population. For example, if females are oversampled and height is a model outcome, $\sum_{i=1}^{n_w} \zeta_i + \sum_{i=1}^{n_m} \zeta_i \neq 0$, with n_w representing the sample of women and n_m the sample of men. This result will occur if gender is not conditioned upon or treated as a moderator in the model. The remedy for this violation seems simple enough; however, the complexity of unequal selection and the model may make it very difficult to condition properly upon all necessary variables. Also, conditioning on sample design variables is not a solution for certain types of models, such as unconditional factor analysis where all observed variables are endogenous.

Unequal probabilities of selection may also bias variance estimates due to sample distributions that have more or less dispersion than the population distribution. The degree of bias in observed variable variances, the diagonal of S, determines the degree of bias in standard error estimates. For intercept and mean standard error estimates, bias in $\bar{z}$ may also come into play. Bias in standard errors due to unequal selection can be upward or downward.

2.5.4 SEM Analysis Using Complex Samples with Unequal Probabilities of Selection

All of the SEM models and notation remain the same, as we have described above, for analysis of complex samples and samples with unequal probabilities of selection. However, depending on how we choose to accommodate the sampling design into the analysis, changes must to be made to the modeling steps, including model specification, estimation, and fit assessment. There are two approaches to accommodating complex samples and unequal probability of selection samples. One approach is to incorporate the sample design features explicitly in models (a model-based approach) and the other approach is to use estimation methods that are robust to the sample design (a design-based approach). Or, it may be desirable to use a combination of these two approaches. The model-based approach requires that the analyst take care at the model specification and respecification steps to incorporate variables related to sample selection. In the design-based approach, the analyst must utilize estimation methods that account for the sample design, including the appropriate model fit statistics at the fit assessment step.

2.5.4.1 Modeling the Sample Design.   Many of the features of complex sample design can be accommodated explicitly in SEM models. Clustering is accounted for in the multilevel models that we present in an earlier section. In these models, the variance components associated with the clusters are disaggregated from the variance associated with observations within clusters. These models also have standard errors for fixed-effects parameters that are robust to clustering. Stratification can also be approximately modeled by including strata variables as fixed, main effects.

As mentioned already, unequal selection probabilities can be adjusted for by conditioning upon the grouping variables used in over- and underselection. For some SEM models, such as regression, tests of whether unequal selection is informative for the model have been developed (DuMouchel & Duncan, 1983; Pfeffermann, 1993). However, these tests have not been extended to the general SEM model.

Modeling the sample design is an ideal approach when it can be done. Modelers generally want to specify the correct model, and the correct model should be robust to sample features such as unequal inclusion of observations. Using a modeling approach also results in more efficient estimation and does not require sample sizes as large as those required for the survey sampling estimation methods described next.

2.5.4.2 Survey Sampling Estimation.   There are many reasons why modeling a complex sample design with unequal probabilities of selection may not be possible. First, we may not be interested in variance components or contextual hypotheses. Or, our clusters of interest may not coincide with the clusters in the sampling process. It is also true that many of the variables required to account for unequal probabilities of selection are unavailable or unknown, this is particularly true for weights that are adjusted for nonresponse where the complete process of unequal inclusion is unknown. Models also become atheoretical and possibly unwieldy when all sample design variables are conditioned upon. There are often simply too many variables to accommodate in a model. It may be necessary to allow many sample design variables to act as moderators, and this is limited by sample and model size. Also, as mentioned previously, even though parameter estimates are consistent, standard errors may still be biased, as a result of complex sample selection procedures. Other issues with regard to modeling with complex samples are addressed in Skinner (1989) and Chambers and Skinner (2003).

Another way to account for complex sample design features is to adjust estimation methods rather than models. This approach is well established in the survey sampling tradition, and the methods are adaptable to SEMs.

Weighted Estimation.   Parameter estimates can be corrected from bias due to unequal probabilities of selection using weighted estimation where the weights are as defined in equation (2.64). For example, weighted estimators for the elements of $\bar{z}$ and S (denoted $\bar{z}_\omega$ and S_ω) can be obtained using the following equations:

$$\bar{x}_\omega = \frac{\sum \omega_i x_i}{\sum \omega_i} \tag{2.68}$$

$$s^2_{xy(\omega)} = \frac{\sum \omega_i \left(x_i - \bar{x}_\omega\right)\left(y_i - \bar{y}_\omega\right)}{\sum \omega_i} \tag{2.69}$$

These estimators are not strictly unbiased, but they are consistent and will be unbiased when $\sum \omega_i$ is a fixed value. Using $\bar{z}_\omega$ and S_ω in the F_{ML} estimator (2.21) would result in consistent parameter estimates. However, subsequent variance estimation of model parameters also requires taking the probability weights into account. Weights can be incorporated into a SEM analysis using standard weighted least squares estimation with weight matrix

$$W = I_n\left(\omega_i\right) \tag{2.70}$$

where I_n is a $n \times n$ square identity matrix. However, the asymptotic covariances matrix will not be correct for probability weights as is described in the next section. Weights may also be incorporated into a *pseudo-maximum likelihood*, P_{ML}, estimator (Skinner, 1989) with likelihood function for the parameter estimates

$$L\left(\theta\right) = \sum_i^n \omega_i \log\left(l_i\right) \tag{2.71}$$

where $L\left(\theta\right)$ involves individual likelihoods such as theose used in the direct ML estimator (2.42). The parameter estimates are obtained by maximizing the $L\left(\theta\right)$, or equivalently by solving the set of score equations

$$U\left(\theta\right) = \frac{\partial L\left(\theta\right)}{\partial\theta} = 0 \tag{2.72}$$

Typically, these are solved using an iterative procedure. The point estimates produced from all three of these methods are consistent under unequal probabilities of selection when the ω_i are the inverse of the probability of inclusion in the sample. Stratification and clustering do not have an effect on parameter estimates whether F_{ML} or P_{ML} is used. These features must be considered, along with the weights, when estimating parameter variance.

Variance Estimation for Stratification, Clustering, and Weighting. Weighted estimators are generally less efficient than unweighted estimators because the variance of the weights themselves adds to the instability of the parameter estimates. Weights also induce heteroscedasticity. It is important to recognize that standard error estimates for probability weighted parameters is different than standard error estimates for variance weighted estimators typically used in WLS analysis for heteroscedasticity. Consider the simple example of the standard error for a weighted mean estimated with normalized frequency or variance weights,

$$\sqrt{\frac{\sum \omega_i \left(y_i - \bar{y}\right)^2}{n\left(n-1\right)}} \tag{2.73}$$

as opposed to the standard error of a weighted mean estimated with normalized probability weights,

$$\sqrt{\frac{\sum \left[\omega_i \left(y_i - \bar{y}\right)\right]^2}{n\left(n-1\right)}} \tag{2.74}$$

The subtle difference is that in equation (2.73) the sum of the weighted squared deviations is computed whereas in equation (2.74), the sum of the weighted deviations squared is computed where the latter typically has more variation. Therefore, simply applying weights in the same manner as variance weights will not produce a correct asymptotic covariance matrix.

Clustering is handled for WLS or P_{ML} with a between-cluster variance estimator, which has been in the survey statistics literature for linear parameters since the 1950's (Hansen, Hurwitz, & Madow, 1953). The between-cluster estimator is unbiased for linear parameters and it allows any nesting structure below nesting within PSU for with-replacement (WR) designs (Williams, 2000). This estimator has been extended for use with nonlinear parameters by using first-order Taylor series linearization to linearize parameters followed by application of the between-cluster variance estimator, also known as the *sandwich estimator* (Binder, 1983). The approximate covariance matrix of estimators for P_{ML}

$$VAR\left(\hat{\theta}\right) = (J)^{-1}\left[\widehat{VAR}\left(\hat{U}\left(\hat{\theta}\right)\right)\right]\left[(J)^{-1}\right]' \qquad (2.75)$$

where $J = \partial\hat{U}\left(\hat{\theta}\right)/\partial\hat{\theta}$ is the weighted sample information matrix, which is the model-based variance estimate, and $\widehat{VAR}\left(\hat{U}\left(\hat{\theta}\right)\right)$ is the design-specific variance of the score equations estimated by linearizing the weighted score equation estimates $\hat{U}\left(\hat{\theta}\right)$ into the linearized variate vector z_{ij} and using the between-cluster variance estimator

$$\widehat{VAR}\left(\hat{U}\left(\hat{\theta}\right)\right) = m\sum_{j}\left(z_j - \bar{z}\right)\left(z_j - \bar{z}\right)'/\left(m-1\right) \qquad (2.76)$$

where $\bar{z}_i$ is the mean within cluster j and m is the total number of clusters.

Stratification can be easily incorporated into (2.76) or any variance estimation simply by calculating variance within the stratum, and then pooling over the strata. Strata are independent by design and therefore simply summing the stratum variances results in total variance.

2.5.5 Future Research

All of the methods established for analyzing complex samples with unequal probabilities of selection in the survey sampling tradition have not been incorporated into SEM software. The estimation techniques described here are primarily for the most common design, which is with-replacement (WR) or approximately WR. Samples that are drawn without-replacement (WOR) require other corrections, including a finite population correction factor. Variance estimation for these samples are also affected by nesting structures that fall within the nesting at the PSU. Other variance estimation methods, such as the jackknife and balanced repeated replication (BRR) methods, could be used to accommodate WOR designs (Wolter, 2007).

The chi-square and other measures of model fit in SEM have not been analyzed thoroughly under conditions of clustered data or weighted estimation. The exact analytical effect on the chi-square, RMSEA, and other fit measures needs to be explicated with further research before conclusions can be made about the model fit statistics including the adjusted Chi-square statistics that are often used with complex samples (Rao & Scott, 1981; Satorra & Bentler, 1994; Yuan & Bentler, 2000). Currently the best methods for performing multiple-degree-of-freedom tests (nested tests) is the Wald test. Another good option is the Bollen–Stine bootstrap (Bollen & Stine, 1992) of the chi-square statistic.

Finally, handling unequal probabilities of selection in multilevel models is a fairly unchartered area. Unequal selection can and often does occur at the level of the cluster as well as the observations within a cluster. Therefore, it is likely that unequal selection will affect estimates at one or more levels of the multilevel model (Pfeffermann et. al. 1998). The correct method for applying probability weights in such a model is relatively unknown, although some work has begun on this (Korn & Graubard, 2003; Kovacević & Rai, 2003; Pfeffermann et al., 1998; Rabe-Hesketh & Skrondal, 2006). The general suggestion is to apply weights at all levels of the model.

2.6 CONCLUSION

The SEM began as a general model that incorporated multiple regression, simultaneous equation models, and factor analysis as special cases. Since its origins, it has continued to grow in its generality and to provide a common framework from which to see typical and less typical techniques in a new light and to extend these models in novel ways. In this chapter, we provided an overview of SEMs and the major steps in implementing them. We also used this framework to show more recent extensions of SEMs to multilevel, mixture, and IRT models and to explain methods by which complex sample designs are taken account of in SEMs. The literature, software, and applications of SEMs are vast, and we provided a selective overview. We highlighted established findings and pointed to areas that require further development. The references cited provide an entry point for those researchers who wish to learn more.

Kenneth A. Bollen
Daniel J. Bauer
Sharon L. Christ
University of North Carolina at Chapel Hill

Michael C. Edwards
The Ohio State University

REFERENCES

Allison, P., & Bollen, K. A. (1997, August). Change score, fixed effects, and random component models: A structural equation approach. Presented at the American Sociological Association Convention. Toronto, Ontario, Canada.

Arminger, G., & Shoenberg, R. J. (1989). Pseudo maximum likelihood estimation and a test for misspecification in mean and covariance structure models. *Psychometrika, 54*, 409–426.

Arminger, G., & Stein, P. (1997). Finite mixtures of covariance structure models with regressors. *Sociological Methods and Research, 26*, 148–182.

Arminger, G., Stein, P., & Wittenberg, J. (1999). Mixtures of conditional mean- and covariance-structure models. *Psychometrika, 64,* 475–494.

Bartholomew, D. J. (1987). *Latent variable models and factor analysis.* London: Griffin.

Bauer, D. J. (2003). Estimating multilevel linear models as structural equation models. *Journal of Educational and Behavioral Statistics, 28,* 135–167.

Bauer, D. J. (2005). A semiparametric approach to modeling nonlinear relations among latent variables. *Structural Equation Modeling: A Multidisciplinary Journal, 4,* 513–535.

Bauer, D. J., & Curran, P. J. (2003). Distributional assumptions of growth mixture models: Implications for over-extraction of latent trajectory classes. *Psychological Methods, 8,* 338–363.

Bauer, D. J., & Curran, P. J. (2004). The integration of continuous and discrete latent variable models: Potential problems and promising opportunities. *Psychological Methods, 9,* 3–29.

Bentler, P. M., & Liang, J. (2003). Two-level mean and covariance structures: Maximum likelihood via an EM algorithm. In S. P. Reise & N. Duan (Eds.), *Multilevel modeling: Methodological advances, issues, and applications* (pp. 53–70). Mahwah, NJ: Lawrence Erlbaum Associates.

Binder, D. (1983). On the variance of asymptotically normal estimators from complex surveys. *International Statistical Review, 51,* 279–292.

Birnbaum, A. (1968). Some latent trait models and their use in inferring an examinee's ability. In F. M. Lord & M. R. Novick (Eds.), *Statistical theories of mental test scores.* Reading, MA: Addison-Wesley.

Blåfield, E. (1980). *Clustering of observations from finite mixtures with structural information.* Jyväskylä, Finland: Jyväskylä University.

Blalock, H. M. (1964). *Causal inference in nonexperimental research.* New York: Norton.

Blalock, H. M. (Ed.). (1971). *Causal models in the social sciences.* Chicago: Aldine.

Bock, R. D., & Aitkin, M. (1981). Marginal maximum likelihood estimation of item parameters: An application of the EM algorithm. *Psychometrika, 46,* 443–459.

Bock, R. D., Gibbons, R., Schilling, S. G., Muraki, E., Wilson, D. T., & Wood, R. (2002). *TESTFACT 4* [Computer software]. Chicago, IL: Scientific Software International, Inc.

Bock, R. D., & Lieberman, M. (1970). Fitting a response model for n dichotomously scored items. *Psychometrika, 35,* 179–197.

Bollen, K. A. (1989). *Structural equations with latent variables.* New York: Wiley.

Bollen, K. A. (1996). An alternative two stage least squares (2SLS) estimator for latent variable equations. *Psychometrika, 61,* 109–121.

Bollen, K. A. (2001). Two-stage least squares and latent variable models: Simultaneous estimation and robustness to misspecifications. In *Structural equation modeling: Present and future* (pp. 119–138). Lincolnwood, IL: Scientific Software International.

Bollen, K. A. (2002). Latent variables in psychology and the social sciences. *Annual Review of Psychology, 53,* 605-634.

Bollen, K. A. (2007). On the origins of latent curve models. In R. Cudeck & R. MacCallum (Eds.), *Factor analysis at 100: Historic developments and future directions* (pp. 79–97). Mahwah, NJ: Lawrence Erlbaum Associates.

Bollen, K. A., & Bauer, D. J. (2004). Automating the selection of modelimplied instrumental variables. *Sociological Methods and Research, 32,* 425–452.

Bollen, K. A., & Long, J. S. (Eds.) (1993). *Testing structural equation models.* Newbury Park, CA: Sage.

Bollen, K. A., & Stine, R. A. (1990). Direct and indirect effects: Classical and bootstrap estimates of variability. *Sociological Methodology, 20,* 115–140.

Bollen, K. A., & Stine, R. A. (1992). Bootstrapping goodness of fit measures in structural equation models. *Sociological Methods and Research, 21,* 205–229.

Bollen, K. A., & Stine, R. A. (1993). Bootstraping goodness-of-fit measures in structural equation models. In K. A. Bollen & J. S. Long (Eds.), *Testing structural equation models* (p. 111–135). Newbury Park, CA: Sage.

Browne, M. W. (1974). Generalized least-squares estimators in the analysis of covariance structures. *South African Statistical Journal, 8,* 1–24.

Browne, M. W. (1984). Asymptotic distribution free methods in analysis of covariance structures. *British Journal of Mathematical and Statistical Psychology, 37,* 62–83.

Browne, M. W., & Zhang, G. (2007). Developments in the factor analysis of individual time series. In R. Cudeck & R. MacCallum (Eds.), *Factor analysis at 100: Historic developments and future directions* (pp. 265–291). Mahwah, NJ: Lawrence Erlbaum Associates.

Cai, L., Maydeu-Olivares, A., Coffman, D. L., & Thissen, D. (2006). Limited-information goodness-of-fit testing of item response theory models for sparse 2p tables. *British Journal of Mathematical and Statistical Psychology, 59,* 173–194.

Chambers, R., & Skinner, C. J. (2003). *Analysis of survey data.* New York: Wiley.

Christoffersson, A. (1975). Factor analysis of dichotomized variables. *Psychometrika, 40,* 5–32.

Cochran, W. G. (1977). *Sampling techniques.* 3rd ed. New York: Wiley.

Curran, P. J. (2003). Have multilevel models been structural equation models all along? *Multivariate Behavioral Research, 38,* 529–569.

Davis, W. R. (1993). The FC1 rule of identification for confirmatory factor analysis: A general sufficient condition. *Sociological Methods and Research, 21,* 403–437.

Dolan, C. V., & van der Maas, H. L. J. (1998). Fitting multivariate normal finite mixtures subject to structural equation modeling. *Psychometrika, 63,* 227–253.

DuMouchel, W. H., & Duncan, G. J. (1983). Using sample survey weights in multiple regression analyses of stratified samples. *Journal of the American Statistical Association, 78,* 535–543.

Duncan, O. D. (1966). Path analysis: Sociological examples. *American Journal of Sociology, 72,* 1–16.

Edwards, M. C. (2005). A Markov chain Monte Carlo approach to confirmatory item factor analysis. Doctoral dissertation, University of North Carolina at Chapel Hill.

Flora, D. B., & Curran, P. J. (2004). An empirical evaluation of alternative methods of estimation for confirmatory factor analysis with ordinal data. *Psychological Methods, 9,* 466–491.

Goldberger, A. S., & Duncan, O. D. (1973). *Structural equation models in the social sciences.* New York: Academic Press.

Goldstein, H. I., & McDonald, R. P. (1988). A general model for the analysis of multilevel data. *Psychometrika, 53,* 455–467.

Hägglund, G. (1982). Factor analysis by instrumental variables. *Psychometrika, 47,* 209–222.

Hansen, M., Hurwitz, W., & Madow, W. (1953). *Sample survey methods and theory.* New York: Wiley.

Hipp, J., & Bauer, D. J. (2006). Local solutions in the estimation of growth mixture models. *Psychological Methods, 11,* 36–53.

Hu, L. T., & Bentler, P. M. (1999). Cutoff criteria for fit indexes in covariance structure analysis: Conventional criteria versus new alternatives. *Structural Equation Modeling, 6,* 1–55.

Jedidi, K., Jagpal, H. S., & DeSarbo, W. S. (1997a). Finite-mixture structural equation models for response-based segmentation and unobserved heterogeneity. *Marketing Science, 16,* 39–59.

Jedidi, K., Jagpal, H. S., & DeSarbo, W. S. (1997b). Stemm: A general finite mixture structural equation model. *Journal of Classification, 14,* 23–50.

Johnston, J. (1984). *Econometric methods.* New York: McGraw-Hill.

Jöreskog, K. G. (1973). A general method for estimating a linear structural equation. In A. Goldberger & O. Duncan (Eds.), *Structural equation models in the social sciences* (pp. 85–112). New York: Academic Press.

Jöreskog, K. G. (1983). Factor analysis as an error-in-variables model. In H. Wainer & S. Messick (Eds.), *Principles of modern psychological measurement* (pp. 185–196). Hillsdale, NJ: Lawrence Erlbaum Associates.

Jöreskog, K. G., & Moustaki, I. (2001). Factor analysis of ordinal variables: A comparison of three approaches. *Multivariate Behavioral Research, 36,* 347–387.

Jöreskog, K. G., & Sörbom, D. (1978). *LISREL IV: Analysis of linear structural relationships by the method of maximum likelihood* [Computer software]. Chicago: National Educational Resources, Inc.

Kish, L. (1965). *Survey sampling.* New York: Wiley.

Korn, E. L., & Graubard, B. I. (2003). Estimating variance components by using survey data. *Journal of Royal Statistical Society, Series A, Statistics in Society, 65,* 175–190.

Kovacević, M. S., & Rai, S. N. (2003). Pseudo maximum likelihood approach to multilevel modelling of survey data. *Communications in Statistics, 32,* 103–121.

Lazarsfeld, P. F., & Henry, N. W. (1968). *Latent structure analysis.* Boston, MA: Houghton Mifflin.

Lee, S. Y. (1990). Multilevel analysis of structural equation models. *Biometrika, 77,* 763–772.

Lee, S. Y., & Poon, W. Y. (1998). Analysis of two-level structural equation models via EM type algorithms. *Statistica Sinica, 8,* 749–766.

Levy, P. S., & Lemeshow, S. (1999). *Sampling of populations: Methods and applications.* 3rd ed. New York: Wiley.

Liang, J. J., & Bentler, P. M. (2004). An em algorithm for fitting two-level structural equation models. *Psychometrika, 69,* 101–122.

Lo, Y., Mendell, N. R., & Rubin, D. B. (2001). Testing the number of components in a normal mixture. *Biometrika, 88,* 767–778.

Lohr, S. L. (1999). *Sampling: Design and analysis.* Pacific Grove, CA: Duxbury Press.

Lord, F. M., & Novick, M. R. (1968). *Statistical theories of mental test scores.* Reading, MA: Addison-Wesley.

Madansky, A. (1964). Instrumental variables in factor analysis. *Psychometrika, 29,* 105–113.

McDonald, R. P., & Goldstein, H. (1989). Balanced versus unbalanced designs for linear structural relations in two-level data. *British Journal of Mathematical and Statistical Psychology, 42,* 215–232.

McLachlan, G. J. (1987). On bootstrapping the likelihood ratio test statistic for the number of components in a normal mixture. *Applied Statistics, 36,* 318–324.

Mehta, P. D., & Neale, M. C. (2005). People are variables too: Multilevel structural equations modeling. *Psychological Methods, 10,* 259–284.

Meredith, W., & Tisak, J. (1984). "Tuckerizing" growth curves. Presented at the annual meeting of the Psychometric Society, Santa Barbara, CA.

Meredith, W., & Tisak, J. (1990). Latent curve analysis. *Psychometrika, 55,* 107–122.

Muthén, B. O. (1978). Contributions to factor analysis of dichotomous variables. *Psychometrika, 43,* 551–560.

Muthén, B. O. (1984). A general structural equation model with dichotomous, ordered categorical, and continuous latent variable indicators. *Psychometrika, 49,* 115–132.

Muthén, B. O. (1989). Latent variable modeling in heterogeneous populations. *Psychometrika, 54,* 557–585.

Muthén, B. O. (1994). Multilevel covariance structure analysis. *Sociological Methods and Research, 22,* 376–398.

Muthén, B. O. (1997). Latent variable modeling with longitudinal and multilevel data. In A. Raftery (Ed.), *Sociological methodology 1997* (pp. 453–480). Washington, DC: American Sociological Association.

Muthén, B. O. (2001). Second-generation structural equation modeling with a combination of categorical and continuous latent variables: New opportunities for latent class/latent growth modeling. In A. Sayer & L. Collins (Eds.), *New methods for the analysis of change,* (pp. 291–322). Washington D.C.: American Psychological Association.

Muthén, B. O. (2003). Statistical and substantive checking in growth mixture modeling: Comment on Bauer and Curran (2003). *Psychological Methods, 8,* 369–377.

Muthén, B. O., du Toit, S. H. C., & Spisic, D. *Robust inference using weighted least squares and quadratic estimating equations in latent variable modeling with categorical and continuous outcomes.* Unpublished manuscript.

Muthén, B. O., & Satorra, A. (1989). Multilevel aspects of varying parameters in structural models. In R. D. Bock (Ed.), *Multilevel analysis of educational data* (pp. 87–99). New York: Academic Press.

Muthén, B. O., & Satorra, A. (1995). Complex sample data in structural equation modeling. In P. Marsden (Ed.), *Sociological methodology 1995* (pp. 216–316). Washington, DC: American Sociological Association.

Muthén, B. O., & Shedden, K. (1999). Finite mixture modeling with mixture outcomes using the EM algorithm. *Biometrics, 55*, 463–469.

Neale, M. C., & Cardon, L. (1992). *Methodology for genetic studies of twins and families.* Nordrecht, The Netherlands: Kluwer Academic.

Neale, M. C., Boker, S. M., Xie, G., & Maes, H. H. (1999). *Mx: Statistical modeling* [computer manual, 5th ed]. Retrieved from http://www.vcu.edu/mx/documentation.html.

Olsson, U. (1979). Maximum likelihood estimation of the polychoric correlation coefficient. *Psychometrika, 44*, 443–460.

Pfeffermann, D. (1993). The role of sampling weights when modeling survey data. *International Statistical Review, 61*, 317–337.

Pfeffermann, D., Skinner, C., Holmes, D., Goldstein, H., & Rasbash, J. (1998). Weighting for unequal selection probabilities in multilevel models. *Journal of the Royal Statistical Society. Series B, Statistical Methodology, 60*, 23–40.

Rabe-Hesketh, S., & Skrondal, A. (2006). Multilevel modelling of complex survey data. *Journal of the Royal Statistical Society, Series A, Statistics in Society, 169*, 805–827.

Rao, J. N. K. and Scott, A. J. (1981). The analysis of categorical data from complex sample surveys: Chi-squared tests for goodness of fit and independence in two-way tables. *The Journal of the American Statistical Association, 76*, 221–230.

Rovine, M. J., & Molenaar, P. C. M. (2000). A structural modeling approach to a multilevel random coefficients model. *Multivariate Behavioral Research, 35*, 51–88.

Rovine, M. J., & Molenaar, P. C. M. (2001). A structural equations modeling approach to the general linear mixed model. In L. M. Collins & A. G. Sayer (Eds.), *New methods for the analysis of change* (pp. 65–96). Washington, DC: American Psychological Association.

Samejima, F. (1969). Estimation of latent ability using a response pattern of graded scores. *Psychometrika Monograph, 17*.

Satorra, A. (1990). Robustness issues in structural equation modeling: A review of recent developments. *Quality and Quantity, 24*, 367–386.

Satorra, A., & Bentler, P. M. (1994). Corrections to test statistics and standard errors in covariance structure analysis. In A. von Eye & C. C. Clogg (Eds.), *Latent variable analysis: Applications to developmental research* (pp. 399–419). Thousand Oaks, CA: Sage.

Sinharay, S. (2005). Assessing fit of unidimensional item response theory models using a Bayesian approach. *Journal of Educational Measurement, 42*, 375–394.

Skinner, C. J. (1989). Domain means, regression and multivariate analysis. In C. J. Skinner, D. Holt, & T. M. F. Smith (Eds.), *Analysis of complex surveys* (pp. 59–87). Wiley: New York.

Skrondal, A., & Rabe-Hesketh, S. (2004). *Generalized latent variable modeling: Multilevel, longitudinal, and structural equation models.* Boca Raton, FL: Chapman & Hall.

Snijders, T., & Bosker, R. (1999). *Multilevel analysis: An introduction to basic and advanced multilevel modeling.* Thousand Oaks, CA: Sage.

Takane, Y., & de Leeuw, J. (1987). On the relationship between item response theory and factor analysis of discretized variables. *Psychometrika, 52*, 393–408.

Titterington, D. M., Smith, A. F. M., & Makov, U. E. (1985). *Statistical analysis of finite mixture distributions*. Chichester, UK: Wiley.

Willett, J. B., & Sayer, A. G. (1994). Using covariance structure analysis to detect correlates and predictors of individual change over time. *Psychological Bulletin, 116*, 363–381.

Williams, R. L. (2000). A note on robust variance estimation for cluster correlated data. *Biometrics, 56*, 645–646.

Wirth, R. J., & Edwards, M. C. (2007). Item factor analysis: Current approaches and future directions. *Psychological Methods, 12*, 58–79.

Wolter, K. (2007). *Introduction to Variance estimation*, 2nd ed. New York: Springer.

Wooldridge, J. M. (2002) *Econometric analysis of cross section and panel data*. Cambridge, MA: MIT Press.

Wothke, W. (1993). Nonpositive Definite Matrices in Structural Modeling. In K. A. Bollen & J. S. Long (Eds.), *Testing structural equation models* (pp. 256–293). Newbury Park, CA: Sage.

Wothke, W. (2000). Longitudinal and multi-group modeling with missing data. In T. D. Little, K. U. Schnabel, & J. Baumert (Eds.), *Modeling longitudinal and multiple group data: Practical issues, applied approaches, and specific examples* (pp. 219–240). Hillsdale, NJ: Lawrence Erlbaum Associates.

Wright, S. (1918). On the nature of size factors. *Genetics, 3*, 367–374.

Wright, S. (1921). Correlation and causation. *Journal of Applied Agricultural Research, 20*, 557–585.

Yuan, K., & Bentler, P. (2000). Three likelihood-based methods for mean and covariance structure analysis with nonnormal missing data. In M. Sobel & M. Becker (Eds.), *Sociological methodology* (pp. 165–200). Washington, D.C.: American Sociological Association.

Yung, Y.-F. (1997). Finite mixtures in confirmatory factor-analysis models. *Psychometrika, 62*, 297–330.

Zhu, H.-T., & Lee, S.-Y. (2001). A Bayesian analysis of finite mixtures in the LISREL model. *Psychometrika, 66*, 133–152.

CHAPTER 3

ORDER-CONSTRAINED PROXIMITY MATRIX REPRESENTATIONS: ULTRAMETRIC GENERALIZATIONS AND CONSTRUCTIONS WITH MATLAB

3.1 INTRODUCTION

A fundamental concept encountered in the field of classification is that of an ultrametric which serves as a mechanism for characterizing collections of hierarchically organized partitions for some given object set, say $S = \{O_1, \ldots, O_n\}$ (see, e.g., the comprehensive discussion in Barthélemy and Guénoche, 1991, Chapter 3). Formally, if $D = \{d_{ij}\}$ is an $n \times n$ matrix, where d_{ij} refers to a measure of dissimilarity for objects O_i and O_j (so implicitly, larger dissimilarities reflect objects that are more dissimilar), D is called an ultrametric (matrix) if the following conditions are satisfied:

(A) $d_{ij} = d_{ji} \geq 0$ for $1 \leq i, j \leq n$ (symmetry and nonnegativity).
(B) $d_{ij} = 0$ if and only if $i = j$ (definiteness).
(C) $d_{ij} \leq \max\{d_{ik}, d_{jk}\}$ for $1 \leq i, j, k \leq n$ (the ultrametric inequality).
Or, equivalently, for any object triple, O_i, O_j, and O_k, the largest two values among d_{ij}, d_{ik}, and d_{jk} are equal.

The key property that provides the means for identifying the collection of partitions induced by D is the ultrametric inequality in (C), and with some notational care, the nonnegativity condition in (A) and/or the definiteness condition in (B) could easily be relaxed without any real loss of generality. Because $\mathbf{D}$ satisfies (C), there are at most T distinct (positive) values in $\mathbf{D}$ (where $T \leq n - 1$), and each distinct value corresponds to a particular partition of the object set S. Specifically, and denoting these distinct values as $d_{(1)} < d_{(2)} < \cdots < d_{(T)}$ (and for completeness, defining $d_{(0)} \equiv 0$), a sequence of partitions, $\mathcal{P}_0, \mathcal{P}_1, \ldots, \mathcal{P}_T$, is produced having, respectively, $C_0 \equiv n > C_1 > \cdots > C_T \equiv 1$ class(es) that satisfies the following four properties listed:

(i) $\mathcal{P}_0$ is the (disjoint) partition of S into $C_0 \equiv n$ classes, where each object forms its own separate class (and all within-class dissimilarities are therefore zero, and trivially less than or equal to $d_{(0)}$).

(ii) $\mathcal{P}_T$ is the (conjoint) partition of S into a single class ($C_T \equiv 1$) containing all n objects, where all within-class dissimilarities are less than or equal to $d_{(T)}$.

(iii) $\mathcal{P}_t$ is the partition of S into C_t classes where all within-class dissimilarities are less than or equal to $d_{(t)}$, and all between-class dissimilarities are strictly greater than $d_{(t)}$.

(iv) For $1 \leq t \leq T$, the classes in $\mathcal{P}_t$ are either present in $\mathcal{P}_{t-1}$ or are formed by uniting two or more classes in $\mathcal{P}_{t-1}$.

The ultrametrics we consider are usually obtained through a process of optimization and by fitting (through least squares) an ultrametric matrix to some originally given $n \times n$ symmetric and nonnegative proximity matrix $\mathbf{P} = \{p_{ij}\}$ ($p_{ij} = p_{ji} \geq 0$ and $p_{ii} = 0$ for $1 \leq i, j \leq n$). Typically, these fitted ultrametrics will have the maximal number of distinct values, so $T = n - 1$, and the one new class in $\mathcal{P}_t$ is formed by uniting a single pair of classes in $\mathcal{P}_{t-1}$. With some abuse of the statistical notion of a parameter (because we commonly search for the partitions in a hierarchy using some type of optimization strategy, along with merely fitting the corresponding ultrametric, we might prefer the use of the less overloaded term weight), one might say for purposes of later comparison that at most $n - 1$ parameters need to be estimated in the construction of a best-fitting ultrametric to a given proximity matrix $\mathbf{P}$.

As should be apparent from the introduction above, strategies for the hierarchical clustering of an object set produce a sequence of nested partitions in which object classes within each successive partition are constructed from the union of classes present at the previous level. In turn, any such sequence of nested partitions can be characterized by an ultrametric, and conversely, any ultrametric generates a nested collection of partitions. There are three major areas of concern in this chapter: (1) In Section 3.2 we discuss the imposition of a given fixed order, or the initial identification of such a constraining order, in constructing and displaying an ultrametric; (2) extensions of the notion of an ultrametric are presented in Section 3.3 to use alternative collections of partitions that are not necessarily nested but which do contain objects within classes consecutive with respect to a particular object ordering. A method for fitting such structures to a given proximity matrix is discussed along with an alternative strategy for graphical representation; (3) for the enhanced visualization of additive trees, in Section 3.4 we develop a rational method of selecting a root by imposing some type of order-constrained representation on the ultrametric component in a decomposition of an additive tree (nonuniquely into an ultrametric and a centroid metric). A simple numerical example is used throughout the chapter based on a data set characterizing the agreement among the Supreme Court justices for the decade of the Rehnquist Court. All the various MATLAB M-files used to illustrate the extensions are available as open-source code from the web site http://cda.psych.uiuc.edu/ordered_reps_mfiles.

3.1.1 Proximity Matrix for Illustration: Agreement Among Supreme Court Justices

On Saturday, July 2, 2005, the lead headline in *The New York Times* read as follows: "O'Connor to Retire, Touching Off Battle Over Court." Opening the story attached to the headline, Richard W. Stevenson wrote, "Justice Sandra Day O'Connor, the first woman to serve on the U.S. Supreme Court and a critical swing vote on abortion and a host of other divisive social issues, announced Friday that she is retiring, setting up a tumultuous fight over her successor." Our interests are in the data set also provided by the *Times* that day, quantifying the (dis)agreement among the Supreme Court justices during the decade they had been together. We give this in Table 3.1 in the form of the percentage of nonunanimous cases in which the justices *dis*agree, from the 1994–95 term through 2003–04 (known as the Rehnquist Court). The dissimilarity matrix (in which larger entries reflect less similar justices) is given in the same row and column order as the *Times* data set, with the justices ordered from "liberal" to "conservative":

    1: John Paul Stevens (St)
    2: Stephen G. Breyer (Br)
    3: Ruth Bader Ginsberg (Gi)
    4: David Souter (So)
    5: Sandra Day O'Connor (Oc)
    6: Anthony M. Kennedy (Ke)
    7: William H. Rehnquist (Re)
    8: Antonin Scalia (Sc)
    9: Clarence Thomas (Th)

For the various illustrations that will come in the sections to follow, we use the Supreme Court data matrix of Table 3.1. It will be loaded into a MATLAB environment with the command 'load supreme_agree.dat'. The supreme_agree.dat file is in simple ASCII form with verbatim contents as follows:

```
.00  .38  .34  .37  .67  .64  .75  .86  .85
.38  .00  .28  .29  .45  .53  .57  .75  .76
.34  .28  .00  .22  .53  .51  .57  .72  .74
.37  .29  .22  .00  .45  .50  .56  .69  .71
.67  .45  .53  .45  .00  .33  .29  .46  .46
.64  .53  .51  .50  .33  .00  .23  .42  .41
.75  .57  .57  .56  .29  .23  .00  .34  .32
.86  .75  .72  .69  .46  .42  .34  .00  .21
```

Table 3.1 Dissimilarities Among the Nine Supreme Court Justices

	St	Br	Gi	So	Oc	Ke	Re	Sc	Th
1 St	.00	.38	.34	.37	.67	.64	.75	.86	.85
2 Br	.38	.00	.28	.29	.45	.53	.57	.75	.76
3 Gi	.34	.28	.00	.22	.53	.51	.57	.72	.74
4 So	.37	.29	.22	.00	.45	.50	.56	.69	.71
5 Oc	.67	.45	.53	.45	.00	.33	.29	.46	.46
6 Ke	.64	.53	.51	.50	.33	.00	.23	.42	.41
7 Re	.75	.57	.57	.56	.29	.23	.00	.34	.32
8 Sc	.86	.75	.72	.69	.46	.42	.34	.00	.21
9 Th	.85	.76	.74	.71	.46	.41	.32	.21	.00

.85 .76 .74 .71 .46 .41 .32 .21 .00

As noted earlier, all of the M-files and data sets we mention are collected at the site http://cda.psych.uiuc.edu/ordered_reps_mfiles and are freely download-loadable from there. So, the various examples we present are easily replicated by the reader (assuming, of course, access to a MATLAB computing environment).

3.2 ORDER-CONSTRAINED ULTRAMETRICS

As one additional fifth characterization of the partition hierarchy induced by a given ultrametric, we have the property: There exists a (nonunique) one-to-one function or permutation, $\rho(\cdot)$, of the n object indices such that each class in the partition $\mathcal{P}_t$, $1 \leq t \leq T$, defines a set of consecutive (or contiguous) objects with respect to the object order $O_{\rho(1)} \prec O_{\rho(2)} \prec \cdots \prec O_{\rho(n)}$. Thus, in forming the new class(es) in $\mathcal{P}_t$ from those in $\mathcal{P}_{t-1}$, only adjacent classes in the ordering may be united. [Some sense of the high degree of nonuniqueness for object orderings that would show the contiguity property for object classes can be developed from the following observation: Suppose that $\rho(\cdot)$ is some permutation providing the contiguity property and consider, for any t, the adjacent classes in $\mathcal{P}_{t-1}$ that are united to form one of the new classes in $\mathcal{P}_t$. A new permutation defined by interchanging the to-be-united adjacent classes but which maintains the same object order within each class would still have the contiguity property for all the object classes in the partition hierarchy.] For any permutation that provides the consecutive order property for the classes of the partitions $\mathcal{P}_0, \ldots, \mathcal{P}_T$, if the rows and columns of the ultrametric matrix $\mathbf{D}$ are reordered by $\rho(\cdot)$ and we define $\mathbf{D}_\rho \equiv \{d_{\rho(i)\rho(j)}\}$, the latter matrix displays what is called an anti-Robinson (AR) form: The patterning of entries in $\mathbf{D}_\rho$ is such that moving away from the main diagonal within any row or any column, the entries never decrease. Formally, the following two order conditions hold:

within rows: $d_{\rho(i)\rho(k)} \leq d_{\rho(i)\rho(j)}$ for $1 \leq i < k < j \leq n$.
within columns: $d_{\rho(k)\rho(j)} \leq d_{\rho(i)\rho(j)}$ for $1 \leq i < k < j \leq n$.

Even more restrictedly, the reordered matrix $\mathbf{D}_\rho$ also shows what is called the *strongly anti-Robinson* (SAR) *form*, which is important for the type of consistent graphical representation we generally hope to provide for a fitted matrix. Specifically, $\mathbf{D}_\rho$ shows the SAR form (see Hubert, Arabie, & Meulman, 1998) because whenever two above-diagonal entries in $\mathbf{D}_\rho$ that are in adjacent columns within the same row are equal, all entries in rows placed earlier for the same two adjacent columns are also equal; similarly, whenever two above-diagonal entries in $\mathbf{D}_\rho$ in adjacent rows within the same column are equal, all entries in columns placed later for the same two adjacent rows are also equal. The importance of an SAR form for constructing consistent graphical representations is discussed in detail in Hubert et al. (1998) along with many historical precedents. For us, we merely note that all of the generalizations pursued in the sections to follow are based on collections of partitions leading invariably to fitted matrices that are SAR. So, all of our representations are automatically SAR by default and provide consistent graphical displays.

Two M-files are discussed below. The first one finds a good least-squares ultrametric that can be displayed consistently with respect to a given constraining input order for the objects that is supplied explicitly by the user (ultrafnd_confit.m); the second (ultrafnd_confnd.m) actually locates a good constraining order to impose in the first place. These two M-files and others discussed later will require several additional M-files associated with the recent text by Hubert, Arabie, and Meulman (2006) (these files are open-source and downloadable separately from http://cda.psych.uiuc.edu/srpm_mfiles; for convenience, the needed M-files are also placed at the earlier ordered_reps_mfiles site as well). We will not spend much time reviewing the Hubert et al. (2006) work but will use it directly in the extensions pursued here. Reference will be made to the demonstrations and discussion in the latter source whenever appropriate.

3.2.1 The M-file ultrafnd_confit.m

The M-file ultrafnd_confit.m serves to identify a good-fitting (in a least-squares sense) ultrametric that could be displayed consistently with respect to a given fixed order. Along with this file, we also provide both its help header comments below and an application to the supreme_agree data. What should be noted is the following: The input proximity matrix (prox) is given as supreme_agree; the permutation that determines the order in which the heuristic optimization strategy seeks the inequality constraints to define the obtained ultrametric is random [the built-in MATLAB function random(9)]; thus, the routine could be rerun to see if local optima are obtained in identifying the ultrametric (but still constrained by exactly the same object order); the constraining permutation (conperm) is given here as the identity (1 : 9). For output, we provide the ultrametric identified in find with variance-accounted-for (VAF) of 73.69%. Generally, a VAF measure is defined as

$$\text{VAF} = 1 - \frac{\sum_{i<j}(p_{ij} - \hat{p}_{ij})^2}{\sum_{i<j}(p_{ij} - \bar{p})^2}$$

where $\hat{p}_{ij}$ represents a fitted proximity (to p_{ij}) and $\bar{p}$ is the mean off-diagonal proximity in $\mathbf{P}$.

For completeness, the best anti-Robinson matrix (least-squares) to the input proximity matrix is given by arobprox with VAF of 99.55%. The found ultrametric would display a VAF of 74.02% compared against this best-fitting anti-Robinson matrix. The reason these later anti-Robinson elements are given is that the order-constrained ultrametric is actually found by imposing an ultrametric on the best-fitting anti-Robinson approximation for the original proximity matrix. Thus, as our computational mechanism for imposing the given ordering on the ultrametric obtained, we simply use the best-fitting anti-Robinson matrix as a point of departure.

```
>> load supreme_agree.dat
>> supreme_agree

supreme_agree =
```

0	0.3800	0.3400	0.3700	0.6700	0.6400	0.7500	0.8600	0.8500
0.3800	0	0.2800	0.2900	0.4500	0.5300	0.5700	0.7500	0.7600
0.3400	0.2800	0	0.2200	0.5300	0.5100	0.5700	0.7200	0.7400
0.3700	0.2900	0.2200	0	0.4500	0.5000	0.5600	0.6900	0.7100
0.6700	0.4500	0.5300	0.4500	0	0.3300	0.2900	0.4600	0.4600
0.6400	0.5300	0.5100	0.5000	0.3300	0	0.2300	0.4200	0.4100
0.7500	0.5700	0.5700	0.5600	0.2900	0.2300	0	0.3400	0.3200
0.8600	0.7500	0.7200	0.6900	0.4600	0.4200	0.3400	0	0.2100
0.8500	0.7600	0.7400	0.7100	0.4600	0.4100	0.3200	0.2100	0

```
>> [find,vaf,vafarob,arobprox,vafultra] = ...
   ultrafnd_confit(supreme_agree,randperm(9),1:9)

find =
```

0	0.3633	0.3633	0.3633	0.6405	0.6405	0.6405	0.6405	0.6405
0.3633	0	0.2850	0.2850	0.6405	0.6405	0.6405	0.6405	0.6405
0.3633	0.2850	0	0.2200	0.6405	0.6405	0.6405	0.6405	0.6405
0.3633	0.2850	0.2200	0	0.6405	0.6405	0.6405	0.6405	0.6405
0.6405	0.6405	0.6405	0.6405	0	0.3100	0.3100	0.4017	0.4017
0.6405	0.6405	0.6405	0.6405	0.3100	0	0.2300	0.4017	0.4017
0.6405	0.6405	0.6405	0.6405	0.3100	0.2300	0	0.4017	0.4017
0.6405	0.6405	0.6405	0.6405	0.4017	0.4017	0.4017	0	0.2100
0.6405	0.6405	0.6405	0.6405	0.4017	0.4017	0.4017	0.2100	0

```
vaf =

    0.7369


vafarob =

    0.9955


arobprox =
```

0	0.3600	0.3600	0.3700	0.6550	0.6550	0.7500	0.8550	0.8550
0.3600	0	0.2800	0.2900	0.4900	0.5300	0.5700	0.7500	0.7600
0.3600	0.2800	0	0.2200	0.4900	0.5100	0.5700	0.7200	0.7400
0.3700	0.2900	0.2200	0	0.4500	0.5000	0.5600	0.6900	0.7100
0.6550	0.4900	0.4900	0.4500	0	0.3100	0.3100	0.4600	0.4600
0.6550	0.5300	0.5100	0.5000	0.3100	0	0.2300	0.4150	0.4150
0.7500	0.5700	0.5700	0.5600	0.3100	0.2300	0	0.3300	0.3300
0.8550	0.7500	0.7200	0.6900	0.4600	0.4150	0.3300	0	0.2100

```
  0.8550    0.7600    0.7400    0.7100    0.4600    0.4150    0.3300    0.2100         0
```

```
vafultra =

   0.7402
```

```
>> help ultrafnd_confit.m
```

```
ULTRAFND_CONFIT finds and fits an ultrametric using iterative projection
heuristically on a symmetric proximity matrix in the $L_{2}$-norm,
constrained by a given object order.

syntax: [find,vaf,vafarob,arobprox,vafultra] = ...
   ultrafnd_confit(prox,inperm,conperm)

PROX is the input proximity matrix (with a zero main diagonal
and a dissimilarity interpretation);
INPERM is a permutation that determines the order in which the
inequality constraints are considered in obtaining the ultrametric;
CONPERM is the given constraining object order;
VAFAROB is the VAF of the anti-Robinson matrix fit, AROBPROX, to PROX;
VAFULTRA is the VAF of the ultrametric fit to AROBPROX;
FIND is the found least-squares matrix (with variance-accounted-for
of VAF) to PROX satisfying the ultrametric constraints, and given
in CONPERM order.
```

3.2.2　The M-file ultrafnd_confnd.m

The M-file ultrafnd_confnd.m carries out a preliminary identification of a good
initial constraining order, and therefore does not require one to be given a priori. The
constraining order (conperm) is now provided as an output vector, and is constructed
by finding a best anti-Robinson matrix fit to the original proximity input matrix (using
methods discussed in Hubert et al. 2006, Chap. 9). We give the help header comments
verbatim below, but because we would be lead to the identity permutation (1:9) as
the constraining order, we do not repeat the same analyses just given.

```
>> help ultrafnd_confnd.m
```

```
ULTRAFND_CONFND finds and fits an ultrametric using iterative projection
heuristically on a symmetric proximity matrix in the $L_{2}$-norm, and
also locates a initial constraining object order.

syntax: [find,vaf,conperm,vafarob,arobprox,vafultra] = ...
   ultrafnd_confnd(prox,inperm)

PROX is the input proximity matrix (with a zero main diagonal
and a dissimilarity interpretation);
INPERM is a permutation that determines the order in which the
inequality constraints are considered in obtaining the ultrametric;
CONPERM is the identified constraining object order;
VAFAROB is the VAF of the anti-Robinson matrix fit, AROBPROX, to PROX;
VAFULTRA is the VAF of the ultrametric fit to AROBPROX;
FIND is the found least-squares matrix (with variance-accounted-for
of VAF) to PROX satisfying the ultrametric constraints, and given
in CONPERM order.
```

3.2.3   Representing an (Order-Constrained) Ultrametric

There are generally three major ways in which an ultrametric might conveniently be displayed. One would be to impose the subdivision structure directly on the fitted matrix with the row and column objects constrained by the given order. The reordered matrix using the row and column order of (St $\prec$ Br $\prec$ Gi $\prec$ So $\prec$ Oc $\prec$ Ke $\prec$ Re $\prec$ Sc $\prec$ Th) is given below; here the blocks of equal-valued entries are highlighted, indicating the partition hierarchy induced by the ultrametric.

	St	Br	Gi	So	Oc	Ke	Re	Sc	Th
St	x	.36	.36	.36	.64	.64	.64	.64	.64
Br	.36	x	.29	.29	.64	.64	.64	.64	.64
Gi	.36	.29	x	.22	.64	.64	.64	.64	.64
So	.36	.29	.22	x	.64	.64	.64	.64	.64
Oc	.64	.64	.64	.64	x	.31	.31	.40	.40
Ke	.64	.64	.64	.64	.31	x	.23	.40	.40
Re	.64	.64	.64	.64	.31	.23	x	.40	.40
Sc	.64	.64	.64	.64	.40	.40	.40	x	.21
Th	.64	.64	.64	.64	.40	.40	.40	.21	x

Second, the sequence of partitions could be provided along with the levels at which they form (i.e., the $n - 1$ distinct values usually making up the ultrametric):

Partition	Level Formed
{{St,Br,Gi,So,Oc,Ke,Re,Sc,Th}}	.64
{{St,Br,Gi,So},{Oc,Ke,Re,Sc,Th}}	.40
{{St,Br,Gi,So},{Oc,Ke,Re},{Sc,Th}}	.36
{{St},{Br,Gi,So},{Oc,Ke,Re},{Sc,Th}}	.31
{{St},{Br,Gi,So},{Oc},{Ke,Re},{Sc,Th}}	.29
{{St},{Br},{Gi,So},{Oc},{Ke,Re},{Sc,Th}}	.23
{{St},{Br},{Gi,So},{Oc},{Ke},{Re},{Sc,Th}}	.22
{{St},{Br},{Gi},{So},{Oc},{Ke},{Re},{Sc,Th}}	.21
{{St},{Br},{Gi},{So},{Oc},{Ke},{Re},{Sc},{Th}}	—

Or third, we could use the graphical method called a dendrogram, shown in Figure 3.1. The objects are arrayed according to the constraining order along a horizontal axis; a vertical axis gives the calibration of when the partitions emerge.

Irrespective of the method of representation, substantively the interpretation remains much the same. There are three very "tight" dyads in {Sc,Th}, {Gi,So}, and {Ke,Re}; {Oc} joins with {Ke,Re}, and {Br} with {Gi,So}, to form, respectively, the "moderate" conservative and liberal clusters. {St} then joins with {Br,Gi,So} to form the liberal left four-object cluster; {Oc,Ke,Re} unites with the dyad of {Sc,Th}

Figure 3.1   Dendrogram (tree) representation for the ordered-constrained ultrametric described in the text (having a VAF of 73.69%).

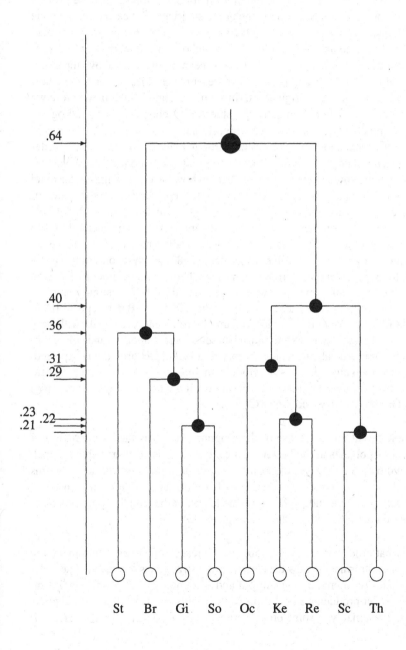

to form the five-object conservative right. In any case, the ordering from left to right used in all three of these representations is from the "far left" to the (in gradual increments) "far right." The dendrogram in Figure 3.1 illustrates well the arbitrariness present in representing such an ultrametric structure graphically. If this object is viewed as a mobile with rotational freedom of movement around the universal joints represented by the solid internal nodes of the diagram, there are 2^8 (and 2^{n-1} in general) equivalent orderings of the nine terminal nodes representing the justices defined by "flipping" the classes around the universal joints. For each terminal node reordering, the pattern of (tied) entries in the fitted ultrametric matrix would remain invariant. So, the particular justice reordering we chose to use (and identified with ultrafnd_confnd.m), has a great deal of extra meaning over and above the structure implicit in the ultrametric imposed. It represents the object order for the best anti-Robinson matrix fit to the original proximity matrix, and is identified before we further impose an ultrametric. Generally, the object order chosen for the dendrogram should place similar objects (according to the original proximities) as close as possible. This is very apparent here, where the particular (identity) constraining order imposed is clearly of the meaningful "liberal" to (gradually) "conservative" variety. There has been quite some interest over the years in how best to arrange the terminal nodes of a dendrogram to produce a more meaningful representation for the encompassed hierarchical clustering. One of the first is from Gruvaeus and Wainer (1972) and implemented in the commercial SYSTAT software package. This method tries to place objects so that those at the edges of a cluster are adjacent to objects outside the cluster to which they are most similar. Degerman (1982) suggests ordering the terminal nodes to maximize a rank correlation measure between node order and a given externally given criterion. Several newer suggestions that are in a similar spirit to these older suggestions include Caraux and Pinloche (2005) and Bar-Joseph, Gifford, and Jaakkola (2001). We note that much of this more recent work deals with very large bioinformatics (e.g., gene expression) data sets where graphical (and color) representations become crucial, and a nonarbitrary terminal object ordering is expected. In this spirit and to again reemphasize why our methods of order-constrained ultrametric representation are important, we end this section with a very nice quote from Parmigiani, Garrett, Irizarry, and Zeger (2003):

An important caveat about the use of dendrograms concerns the interpretation of the order of objects at the bottom. At each split, it is arbitrary which branch is drawn on the left versus the right. As a result, a multitude of dendrograms and orderings are consistent with a given hierarchical classification. Closeness of objects should be judged based on the length of the path that connects them and not on their distance in the ordering. (p. 25)

In short, what order-constrained ultrametric representation does is to remove the high degree of arbitrariness in how the classification structure is displayed. The order of terminal nodes *is* commonly meaningful and if found by ultrafnd_confnd.m, represents the order producing a best-fitting anti-Robinson form to the original proximity matrix. Presumably, if some order is imposed through ultrafnd_confit.m, it will have some substantively defensible interpretation or it should not have been used to begin with.

3.2.4 Alternative (and Generalizable) Graphical Representation for an Ultrametric

Two rather distinct graphical ways for displaying an ultrametric are given in Figures 3.1 and 3.2. Figure 3.1 is in the form of a traditional dendrogram (or a graph-theoretic tree) where the distinct ultrametric values are used to calibrate the vertical axis and indicate the level at which two classes are united to form a new class in a partition within the hierarchy. Each new class formed is represented by a closed circle and is referred to as an *internal node* of the tree. Considering the nine justices to be the terminal nodes (represented by open circles and listed left to right in the constraining order), the ultrametric value between any two objects can also be constructed by taking one-half of the minimum path length between the two corresponding terminal nodes (proceeding upward from one terminal node through the internal node that defines the first class in the partition hierarchy containing them both, and then back down to the other terminal node, with all horizontal lengths in the tree used for graphical purposes only and assumed to be of length zero). Or if the vertical axis calibrations were themselves halved, the minimum path lengths would provide the fitted ultrametric values directly. There is one distinguished node in the tree of Figure 3.1 (indicated by the biggest solid circle), referred to as the *root*, with the property of being equidistant from all terminal nodes. In contrast to various additive tree representations that we give later, the defining characteristic for an ultrametric is that there does exist a position on the tree equidistant from all terminal nodes.

Figure 3.2 provides an alternative representation for an ultrametric (that will prove useful later for ultrametric extensions). Here, a partition is characterized by a set of horizontal lines that each encompass the objects in a particular class. This presentation is possible because the justices are listed from left to right in the same order as that used to constrain the construction of the ultrametric, and thus, each class of a partition contains objects contiguous with respect to this ordering. The calibration on the vertical axis next to each set of horizontal lines representing a specific partition is the increment to the fitted dissimilarity between two particular justices if that pair is *not* encompassed by a continuous horizontal line for a class in this partition. For an ultrametric, a nonnegative increment value for the partition $\mathcal{P}_t$ is just $d_{(t+1)} - d_{(t)} \geq 0$ for $0 \leq t \leq T - 1$ (letting $d_{(0)} \equiv 0$, and noting that an increment for the trivial partition containing a single class, $\mathcal{P}_T$, is not defined nor given in the representation of Figure 3.2). As an example and considering the pair (Oc,Sc), horizontal lines do not encompass this pair except for the last (nontrivial) partition $\mathcal{P}_7$; thus, the fitted ultrametric value of .40 is the sum of the increments attached to the partitions $\mathcal{P}_0, \ldots, \mathcal{P}_6$: $.21 + .01 + .01 + .06 + .02 + .05 + .04 = .40$.

Figure 3.2   Alternative representation for the fitted values of the order-constrained ultrametric (having a VAF of 73.69%).

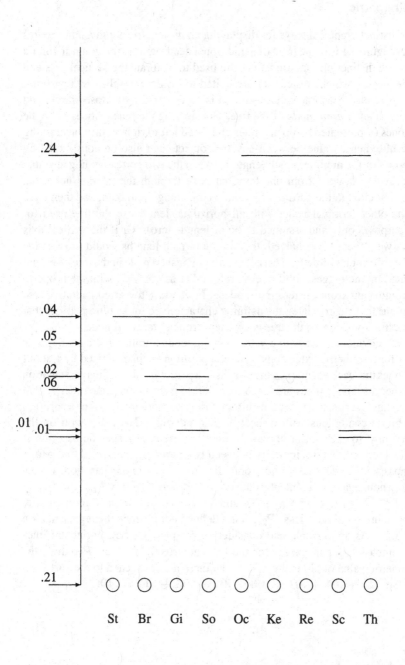

3.2.5 Alternative View of Ultrametric Matrix Decomposition

A general mechanism exists for decomposing any ultrametric matrix $\mathbf{D}$ into a (nonnegatively) weighted sum of dichotomous (0/1) matrices, each representing one of the partitions of the hierarchy, $\mathcal{P}_0, \ldots, \mathcal{P}_T$, induced by $\mathbf{D}$. Specifically, if $\mathbf{P}_t = \{p_{ij}^{(t)}\}$, for $0 \leq t \leq T - 1$, is an $n \times n$ symmetric 0/1 dissimilarity matrix corresponding to $\mathcal{P}_t$ in which an entry $p_{ij}^{(t)}$ is 0 if O_i and O_j belong to the same class in $\mathcal{P}_t$ and is otherwise equal to 1, then for some collection of suitably chosen nonnegative weights, $\alpha_0, \alpha_1, \ldots, \alpha_{T-1}$,

$$\mathbf{D} = \sum_{t=0}^{T-1} \alpha_t \mathbf{P}_t$$

Generally, the nonnegative weights $\alpha_0, \alpha_1, \ldots, \alpha_{T-1}$ are given by the (differences in) partition increments that calibrate the vertical axis of the dendrogram. Moreover, because the ultrametric represented by Figure 3.1 was generated by optimizing a least-squares loss function in relation to a given proximity matrix $\mathbf{P}$, an alternative interpretation for the weights obtained is that they solve the nonnegative least-squares task of

$$\min_{\{\alpha_t \geq 0, \ 0 \leq t \leq T-1\}} \sum_{i<j} \left[p_{ij} - \sum_{t=0}^{T-1} \alpha_t p_{ij}^{(t)} \right]^2 \tag{3.1}$$

for the fixed collection of dichotomous matrices $\mathbf{P}_0, \mathbf{P}_1, \ldots, \mathbf{P}_{T-1}$. Although the solution to (1) is generated indirectly in this case from the least-squares optimal ultrametric fitted directly to $\mathbf{P}$, in general (and especially for the extensions we pursue), for any fixed proximity matrix $\mathbf{P}$ and collection of dichotomous matrices $\mathbf{P}_0, \ldots, \mathbf{P}_{T-1}$, however obtained, the nonnegative weights α_t, $0 \leq t \leq T - 1$, that solve (1) can be obtained with any nonnegative least-squares method. We will routinely use in particular (and without further comment) the code rewritten in MATLAB for a subroutine originally provided by Wollan and Dykstra (1987, pp. 238–240) based on a strategy for solving linear inequality constrained least-squares tasks by iterative projection (see also Dykstra, 1983).

In the verbatim script below, we show how the M-file `partitionfit.m` can be used to reconstruct the order-constrained ultrametric of Section 2.3. The crucial component is in constructing the $m \times n$ matrix (member) that defines class membership for the $m = 8$ nontrivial partitions generating the ultrametric. Note in particular that we do not include the unnecessary conjoint partition involving a single class (in fact, its inclusion would produce a numerical error in the Wollan and Dykstra least-squares subcode integral to `partitionfit.m`; thus, there would be a nonzero `end_condition` value). The M-file `partitionfit.m` will be relied upon again in the sections to follow to generalize the type of structural representations possible for a proximity matrix beyond that of just an order-constrained ultrametric.

```
>> help partitionfit.m

  PARTITIONFIT provides a least-squares approximation to a proximity
```

matrix based on a given collection of partitions.

syntax: [fitted,vaf,weights,end_condition] = partitionfit(prox,member)

PROX is the n x n input proximity matrix (with a zero main diagonal
and a dissimilarity interpretation); MEMBER is the m x n matrix
indicating cluster membership, where each row corresponds to a specific
partition (there are m partitions in general); the columns of MEMBER
are in the same input order used for PROX.
FITTED is an n x n matrix fitted to PROX (through least-squares)
constructed from the nonnegative weights given in the m x 1 WEIGHTS
vector corresponding to each of the partitions. VAF is the variance-
accounted-for in the proximity matrix PROX by the fitted matrix FITTED.
END_CONDITION should be zero for a normal termination of the optimization
process.

```
>> member = [1 1 1 1 2 2 2 2 2;1 1 1 1 2 2 2 3 3;1 2 2 2 3 3 3 4 4;1 2 2 2 3 4 4 5 5;
1 2 3 3 4 5 5 6 6;1 2 3 3 4 5 6 7 7;1 2 3 4 5 6 7 8 8;1 2 3 4 5 6 7 8 9]

member =

    1    1    1    1    2    2    2    2    2
    1    1    1    1    2    2    2    3    3
    1    2    2    2    3    3    3    4    4
    1    2    2    2    3    4    4    5    5
    1    2    3    3    4    5    5    6    6
    1    2    3    3    4    5    6    7    7
    1    2    3    4    5    6    7    8    8
    1    2    3    4    5    6    7    8    9

>> [fitted,vaf,weights,end_condition] = partitionfit(supreme_agree,member)

fitted =

        0   0.3633   0.3633   0.3633   0.6405   0.6405   0.6405   0.6405   0.6405
   0.3633        0   0.2850   0.2850   0.6405   0.6405   0.6405   0.6405   0.6405
   0.3633   0.2850        0   0.2200   0.6405   0.6405   0.6405   0.6405   0.6405
   0.3633   0.2850   0.2200        0   0.6405   0.6405   0.6405   0.6405   0.6405
   0.6405   0.6405   0.6405   0.6405        0   0.3100   0.3100   0.4017   0.4017
   0.6405   0.6405   0.6405   0.6405   0.3100        0   0.2300   0.4017   0.4017
   0.6405   0.6405   0.6405   0.6405   0.3100   0.2300        0   0.4017   0.4017
   0.6405   0.6405   0.6405   0.6405   0.4017   0.4017   0.4017        0   0.2100
   0.6405   0.6405   0.6405   0.6405   0.4017   0.4017   0.4017   0.2100        0

vaf =

   0.7369

weights =

   0.2388
   0.0383
   0.0533
   0.0250
   0.0550
   0.0100
   0.0100
   0.2100

end_condition =

   0
```

3.3 ULTRAMETRIC EXTENSIONS BY FITTING PARTITIONS CONTAINING CONTIGUOUS SUBSETS

The M-file `partitionfit.m` of Section 3.2 is a very general routine giving a least-squares approximation to a proximity matrix based on a given collection of partitions. Thus, no matter how the set of candidate partitions might be chosen, a least-squares fitted matrix to the given proximity matrix is achieved. For example, if we simply use the nested partitions constructed from an ultrametric, the ultrametric would be retrieved when the latter is used as the input proximity matrix. In this section we show how `partitionfit.m` can also be used to select partitions from a predefined set (this selection is done by those partitions assigned nonnegative weights) that might serve to reconstruct the proximity matrix well. The M-file `consec_subsetfit.m` defines $(n(n-1)/2) - 1$ candidate partitions each characterized by a single contiguous cluster of objects, with all objects before and after this contiguous set forming individual clusters of the partition [the minus 1 appears in the count since the (conjoint) partition defined by a single contiguous set is excluded]. The M-file `consec_subsetfit_alter.m` varies the specific definition of the partitions by including all objects before and all objects after the contiguous set (when nonempty) in forming separate individual clusters of the partitions.

As can be seen from the verbatim output provided below, the nonnegative weighted partitions from `consec_subsetfit.m`, producing a fitted matrix with a VAF of 92.61% are as follows:

Partition	Partition Increment
{{St,Br,Gi,So},{Oc},{Ke},{Re},{Sc},{Th}}	.1939
{{St,Br,Gi,So,Oc},{Ke},{Re},{Sc},{Th}}	.0300
{{St,Br,Gi,So,Oc,Ke},{Re},{Sc},{Th}}	.0389
{{St,Br,Gi,So,Oc,Ke,Re},{Sc},{Th}}	.1315
{{St},{Br,Gi,So,Oc,Ke,Re,Sc,Th}}	.1152
{{St},{Br},{Gi,So,Oc,Ke,Re,Sc,Th}}	.0052
{{St},{Br},{Gi},{So,Oc,Ke,Re,Sc,Th}}	.0153
{{St},{Br},{Gi},{So},{Oc,Ke,Re,Sc,Th}}	.2220
{{St},{Br},{Gi},{So},{Oc},{Ke,Re,Sc,Th}}	.0633
{{St},{Br},{Gi},{So},{Oc},{Ke},{Re,Sc,Th}}	.0030

Similarly, we have a very high VAF of 98.12% based on the more numerous partitions generated from `consec_subsetfit_alter.m`:

Partition	Partition Increment
{{St,Br},{Gi,So,Oc,Ke,Re,Sc,Th}}	.0021
{{St,Br,Gi},{So,Oc,Ke,Re,Sc,Th}}	.0001
{{St,Br,Gi,So},{Oc,Ke,Re,Sc,Th}}	.0001
{{St,Br,Gi,So,Oc,Ke},{Re,Sc,Th}}	.0100
{{St,Br,Gi,So,Oc,Ke,Re},{Sc,Th}}	.1218
{{St},{Br,Gi},{So,Oc,Ke,Re,Sc,Th}}	.0034
{{St},{Br,Gi,So,Oc},{Ke,Re,Sc,Th}}	.0056
{{St},{Br,Gi,So,Oc,Ke,Re},{Sc,Th}}	.0113
{{St},{Br,Gi,So,Oc,Ke,Re,Sc},{Th}}	.0038
{{St},{Br,Gi,So,Oc,Ke,Re,Sc,Th}}	.1170
{{St,Br},{Gi,So},{Oc,Ke,Re,Sc,Th}}	.0165
{{St,Br},{Gi,So,Oc,Ke,Re,Sc,Th}}	.0095
{{St,Br,Gi},{So,Oc},{Ke,Re,Sc,Th}}	.0197
{{St,Br,Gi},{So,Oc,Ke,Re,Sc,Th}}	.0115
{{St,Br,Gi,So},{Oc,Ke,Re,Sc,Th}}	.2294
{{St,Br,Gi,So,Oc},{Ke,Re,Sc,Th}}	.0353
{{St,Br,Gi,So,Oc,Ke},{Re,Sc,Th}}	.0400
{{St,Br,Gi,So,Oc,Ke,Re},{Sc,Th}}	.0132
{{St},{Br},{Gi},{So},{Oc},{Ke},{Re},{Sc},{Th}}	.2050

```
>> help consec_subsetfit.m

   CONSEC_SUBSETFIT defines a collection of partitions involving
   consecutive subsets for the object set and then calls partitionfit.m
   to fit a least-squares approximation to the input proximity matrix based
   on these identified partitions.

   syntax [fitted,vaf,weights,end_condition,member] = consec_subsetfit(prox)

   PROX is the n x n input proximity matrix (with a zero main diagonal
   and a dissimilarity interpretation); MEMBER is the m x n matrix
   indicating cluster membership, where each row corresponds to a specific
   partition (there are m partitions in general); the columns of MEMBER
   are in the same input order used for PROX. The partitions are defined
   by a single contiguous cluster of objects, with all objects before and
   after this contiguous set forming individual clusters of the partitions.
   The value of m is (n*(n-1)/2) - 1; the partition defined by a single
   contiguous partition is excluded.
   FITTED is an n x n matrix fitted to PROX (through least-squares)
   constructed from the nonnegative weights given in the m x 1 WEIGHTS
   vector corresponding to each of the partitions. VAF is the variance-
   accounted-for in the proximity matrix PROX by the fitted matrix FITTED.
   END_CONDITION should be zero for a normal termination of the optimization
   process.

>> load supreme_agree.dat
>> [fitted,vaf,weights,end_condition,member] = consec_subsetfit(supreme_agree);
>> fitted

fitted =

        0    0.4239    0.4239    0.4239    0.6178    0.6478    0.6866    0.8181    0.8181
   0.4239         0    0.3087    0.3087    0.5026    0.5326    0.5715    0.7029    0.7029
   0.4239    0.3087         0    0.3035    0.4974    0.5274    0.5663    0.6977    0.6977
   0.4239    0.3087    0.3035         0    0.4821    0.5121    0.5510    0.6824    0.6824
   0.6178    0.5026    0.4974    0.4821         0    0.2901    0.3290    0.4604    0.4604
```

```
    0.6478    0.5326    0.5274    0.5121    0.2901         0    0.2657    0.3972    0.3972
    0.6866    0.5715    0.5663'   0.5510    0.3290    0.2657         0    0.3942    0.3942
    0.8181    0.7029    0.6977    0.6824    0.4604    0.3972    0.3942         0    0.3942
    0.8181    0.7029    0.6977    0.6824    0.4604    0.3972    0.3942    0.3942         0

>> vaf

vaf =

    0.9261

>> weights

weights =

         0
         0
    0.1939
    0.0300
    0.0389
    0.1315
         0
         0
         0
         0
         0
         0
         0
    0.1152
         0
         0
         0
         0
         0
    0.0052
         0
         0
         0
         0
    0.0153
         0
         0
         0
    0.2220
         0
         0
    0.0633
         0
    0.0030
         0
         0

>> end_condition

end_condition =

     0

>> member

member =

     1     1     3     4     5     6     7     8     9
     1     1     1     4     5     6     7     8     9
     1     1     1     1     5     6     7     8     9
     1     1     1     1     1     6     7     8     9
     1     1     1     1     1     1     7     8     9
```

```
1   1   1   1   1   1   1   8   9
1   1   1   1   1   1   1   1   9
1   2   2   4   5   6   7   8   9
1   2   2   2   5   6   7   8   9
1   2   2   2   2   6   7   8   9
1   2   2   2   2   2   7   8   9
1   2   2   2   2   2   2   8   9
1   2   2   2   2   2   2   2   9
1   2   2   2   2   2   2   2   2
1   2   3   3   5   6   7   8   9
1   2   3   3   3   6   7   8   9
1   2   3   3   3   3   7   8   9
1   2   3   3   3   3   3   8   9
1   2   3   3   3   3   3   3   9
1   2   3   3   3   3   3   3   3
1   2   3   4   4   6   7   8   9
1   2   3   4   4   4   7   8   9
1   2   3   4   4   4   4   8   9
1   2   3   4   4   4   4   4   9
1   2   3   4   4   4   4   4   4
1   2   3   4   5   5   7   8   9
1   2   3   4   5   5   5   8   9
1   2   3   4   5   5   5   5   9
1   2   3   4   5   5   5   5   5
1   2   3   4   5   6   6   8   9
1   2   3   4   5   6   6   6   9
1   2   3   4   5   6   6   6   6
1   2   3   4   5   6   7   7   9
1   2   3   4   5   6   7   7   7
1   2   3   4   5   6   7   8   8
1   2   3   4   5   6   7   8   9
```

```
>> help consec_subsetfit_alter.m

    CONSEC_SUBSETFIT_ALTER defines a collection of partitions involving
    consecutive subsets for the object set and then calls partitionfit.m
    to fit a least-squares approximation to the input proximity matrix based
    on these identified partitions.

    syntax [fitted,vaf,weights,end_condition,member] = ...
                        consec_subsetfit_alter(prox)

    PROX is the n x n input proximity matrix (with a zero main diagonal
    and a dissimilarity interpretation); MEMBER is the m x n matrix
    indicating cluster membership, where each row corresponds to a specific
    partition (there are m partitions in general); the columns of MEMBER
    are in the same input order used for PROX. The partitions are defined
    by a single contiguous cluster of objects, with all objects before and
    all objects after this contiguous set (when nonempty) forming
    separate individual clusters of the partitions.
    (These possible three-class partitions when before and after subsets are
    both nonempty) distinguish consec_subsetfit_alter.m from consec_subsetfit.m).
    The value of m is (n*(n-1)/2) - 1; the partition defined by a single
    contiguous partition is excluded.
    FITTED is an n x n matrix fitted to PROX (through least-squares)
    constructed from the nonnegative weights given in the m x 1 WEIGHTS
    vector corresponding to each of the partitions.  VAF is the variance-
    accounted-for in the proximity matrix PROX by the fitted matrix FITTED.
    END_CONDITION should be zero for a normal termination of the optimization
    process.

>> [fitted,vaf,weights,end_condition,member] = consec_subsetfit_alter(supreme_agree)

fitted =

        0   0.3460   0.3740   0.4053   0.6347   0.6700   0.7200   0.8550   0.8550
   0.3460        0   0.2330   0.2677   0.4971   0.5380   0.5880   0.7342   0.7380
```

0.3740	0.2330	0	0.2396	0.4855	0.5264	0.5764	0.7227	0.7264
0.4053	0.2677	0.2396	0	0.4509	0.5114	0.5614	0.7076	0.7114
0.6347	0.4971	0.4855	0.4509	0	0.2655	0.3155	0.4617	0.4655
0.6700	0.5380	0.5264	0.5114	0.2655	0	0.2550	0.4012	0.4050
0.7200	0.5880	0.5764	0.5614	0.3155	0.2550	0	0.3512	0.3550
0.8550	0.7342	0.7227	0.7076	0.4617	0.4012	0.3512	0	0.2087
0.8550	0.7380	0.7264	0.7114	0.4655	0.4050	0.3550	0.2087	0

vaf =

   0.9812

weights =

   0.0021
   0.0001
   0.0001
        0
   0.0100
   0.1218
        0
   0.0034
        0
   0.0056
        0
   0.0113
   0.0038
   0.1170
   0.0165
        0
        0
        0
        0
   0.0095
   0.0197
        0
        0
        0
   0.0115
        0
        0
        0
   0.2294
        0
        0
   0.0353
        0
   0.0400
   0.0132
   0.2050

end_condition =

   0

member =

1	1	9	9	9	9	9	9	9
1	1	1	9	9	9	9	9	9
1	1	1	1	9	9	9	9	9
1	1	1	1	1	9	9	9	9
1	1	1	1	1	1	9	9	9
1	1	1	1	1	1	1	9	9

```
1   1   1   1   1   1   1   1   9
1   2   2   9   9   9   9   9   9
1   2   2   2   9   9   9   9   9
1   2   2   2   2   9   9   9   9
1   2   2   2   2   2   9   9   9
1   2   2   2   2   2   2   9   9
1   2   2   2   2   2   2   2   9
1   2   2   2   2   2   2   2   2
1   1   3   3   9   9   9   9   9
1   1   3   3   3   9   9   9   9
1   1   3   3   3   3   9   9   9
1   1   3   3   3   3   3   9   9
1   1   3   3   3   3   3   3   9
1   1   3   3   3   3   3   3   3
1   1   1   4   4   9   9   9   9
1   1   1   4   4   4   9   9   9
1   1   1   4   4   4   4   9   9
1   1   1   4   4   4   4   4   9
1   1   1   4   4   4   4   4   4
1   1   1   1   5   5   9   9   9
1   1   1   1   5   5   5   9   9
1   1   1   1   5   5   5   5   9
1   1   1   1   5   5   5   5   5
1   1   1   1   1   6   6   9   9
1   1   1   1   1   6   6   6   9
1   1   1   1   1   6   6   6   6
1   1   1   1   1   1   7   7   9
1   1   1   1   1   1   7   7   7
1   1   1   1   1   1   1   8   8
1   2   3   4   5   6   7   8   9
```

To see how well we might do in relation to choosing only eight partitions to consider (i.e., the same number defining the order-constrained best-fitting ultrametric), we chose in both instances the single (disjoint) partition defined by nine separate classes plus seven partitions that have the highest assigned weights. For those picked from the partition pool identified by consec_subsetfit.m, the VAF drops slightly from 92.61% to 92.51% using the partitions

Partition	Partition Increment
{{St,Br,Gi,So},{Oc},{Ke},{Re},{Sc},{Th}}	.1923
{{St,Br,Gi,So,Oc},{Ke},{Re},{Sc},{Th}}	.0301
{{St,Br,Gi,So,Oc,Ke},{Re},{Sc},{Th}}	.0396
{{St,Br,Gi,So,Oc,Ke,Re},{Sc},{Th}}	.1316
{{St},{Br,Gi,So,Oc,Ke,Re,Sc,Th}}	.1224
{{St},{Br},{Gi},{So},{Oc,Ke,Re,Sc,Th}}	.2250
{{St},{Br},{Gi},{So},{Oc},{Ke,Re,Sc,Th}}	.0671
{{St},{Br},{Gi},{So},{Oc},{Ke},{Re},{Sc},{Th}}	.0000

For those selected from the set generated by consec_subsetfit_alter.m, the VAF again drops slightly, from 98.12% to 97.97%. But in some absolute sense given the size of the VAF, the eight partitions listed below seem to be about all that can be extracted from this particular justice data set.

Partition	Partition Increment
{{St,Br,Gi,So,Oc,Ke,Re},{Sc,Th}}	.1466
{{St},{Br,Gi,So,Oc,Ke,Re,Sc,Th}}	.1399
{{St,Br},{Gi,So},{Oc,Ke,Re,Sc,Th}}	.0287
{{St,Br,Gi},{So,Oc},{Ke,Re,Sc,Th}}	.0326
{{St,Br,Gi,So},{Oc,Ke,Re,Sc,Th}}	.2269
{{St,Br,Gi,So,Oc},{Ke,Re,Sc,Th}}	.0316
{{St,Br,Gi,So,Oc,Ke},{Re,Sc,Th}}	.0500
{{St},{Br},{Gi},{So},{Oc},{Ke},{Re},{Sc},{Th}}	.2051

The three highest weighted partitions have very clear interpretations: {Sc,Th} versus the rest; {St} versus the rest; {St,Br,Gi,So} as the left versus {Oc,Ke,Re,Sc,Th} as the right. The few remaining partitions revolve around several other less salient (adjacent) object pairings that are also very interpretable in relation to the object ordering from liberal to conservative. We give a graphical representation of the latter culled collection of partitions in Figure 3.3. Again, the partition increments are not included in a fitted value whenever a continuous horizontal line encompasses the relevant objects in a defining cluster of the partition.

```
>> member = [1 1 1 1 5 6 7 8 9;1 1 1 1 6 7 8 9;1 1 1 1 1 7 8 9;1 1 1 1 1 1 8 9;
1 2 2 2 2 2 2 2;1 2 3 4 5 5 5 5;1 2 3 4 5 6 6 6;1 2 3 4 5 6 7 8 9]

member =

    1    1    1    1    5    6    7    8    9
    1    1    1    1    1    6    7    8    9
    1    1    1    1    1    1    7    8    9
    1    1    1    1    1    1    1    8    9
    1    2    2    2    2    2    2    2    2
    1    2    3    4    5    5    5    5    5
    1    2    3    4    5    6    6    6    6
    1    2    3    4    5    6    7    8    9

>> [fitted,vaf,weights,end_condition] = partitionfit(supreme_agree,member)

fitted =

         0    0.4245    0.4245    0.4245    0.6168    0.6469    0.6865    0.8181    0.8181
    0.4245         0    0.3021    0.3021    0.4944    0.5245    0.5641    0.6957    0.6957
    0.4245    0.3021         0    0.3021    0.4944    0.5245    0.5641    0.6957    0.6957
    0.4245    0.3021    0.3021         0    0.4944    0.5245    0.5641    0.6957    0.6957
    0.6168    0.4944    0.4944    0.4944         0    0.2895    0.3291    0.4607    0.4607
    0.6469    0.5245    0.5245    0.5245    0.2895         0    0.2620    0.3936    0.3936
    0.6865    0.5641    0.5641    0.5641    0.3291    0.2620         0    0.3936    0.3936
    0.8181    0.6957    0.6957    0.6957    0.4607    0.3936    0.3936         0    0.3936
    0.8181    0.6957    0.6957    0.6957    0.4607    0.3936    0.3936    0.3936         0

vaf =

    0.9251

weights =

    0.1923
```

Figure 3.3 Representation for the fitted values of the (generalized) structure described in the text (having a VAF of 97.97%).

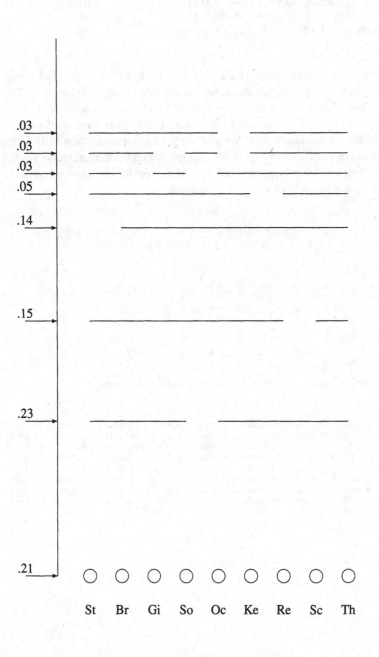

```
     0.0301
     0.0396
     0.1316
     0.1224
     0.2350
     0.0671
          0


end_condition =

     0

>> member = [1 1 1 1 1 1 1 9 9;1 2 2 2 2 2 2 2 2;1 1 3 3 9 9 9 9 9;1 1 1 4 4 9 9 9 9;
1 1 1 1 5 5 5 5 5;1 1 1 1 1 6 6 6 6;1 1 1 1 1 1 7 7 7;1 2 3 4 5 6 7 8 9]

member =

     1     1     1     1     1     1     1     9     9
     1     2     2     2     2     2     2     2     2
     1     1     3     3     9     9     9     9     9
     1     1     1     4     4     9     9     9     9
     1     1     1     1     5     5     5     5     5
     1     1     1     1     1     6     6     6     6
     1     1     1     1     1     1     7     7     7
     1     2     3     4     5     6     7     8     9

>> [fitted,vaf,weights,end_condition] = partitionfit(supreme_agree,member)

fitted =

          0    0.3450    0.3736    0.4062    0.6331    0.6647    0.7147    0.8613    0.8613
     0.3450         0    0.2337    0.2664    0.4933    0.5248    0.5748    0.7215    0.7215
     0.3736    0.2337         0    0.2377    0.4933    0.5248    0.5748    0.7215    0.7215
     0.4062    0.2664    0.2377         0    0.4606    0.5248    0.5748    0.7215    0.7215
     0.6331    0.4933    0.4933    0.4606         0    0.2693    0.3193    0.4659    0.4659
     0.6647    0.5248    0.5248    0.5248    0.2693         0    0.2551    0.4017    0.4017
     0.7147    0.5748    0.5748    0.5748    0.3193    0.2551         0    0.3517    0.3517
     0.8613    0.7215    0.7215    0.7215    0.4659    0.4017    0.3517         0    0.2051
     0.8613    0.7215    0.7215    0.7215    0.4659    0.4017    0.3517    0.2051         0


vaf =

     0.9797


weights =

     0.1466
     0.1399
     0.0287
     0.0326
     0.2269
     0.0316
     0.0500
     0.2051


end_condition =

     0
```

3.3.1 Ordered Partition Generalizations

Given the broad characterization of the properties of an ultrametric described earlier, the generalization mentioned in this subsection rests on merely altering the type of partition allowed in the sequence $\mathcal{P}_0, \mathcal{P}_1, \ldots, \mathcal{P}_T$. Specifically, we will still use an object order $O_{\rho(1)} \prec \cdots \prec O_{\rho(n)}$, and a collection of partitions with fewer and fewer classes consistent with this order by requiring the classes within each partition to contain contiguous objects. However, the constraint will be removed that the new classes in $\mathcal{P}_t$ be formed by uniting only existing classes in $\mathcal{P}_{t-1}$. Thus, although class contiguity is maintained with respect to the same object order in the partitions we identify, the requirement that the classes be nested is relaxed so that if a class is present in $\mathcal{P}_{t-1}$, it will no longer need to appear either as a class by itself or be properly contained within some class in $\mathcal{P}_t$. In comparison to an ultrametric, the extension pursued also requires no more than $n - 1$ estimated parameters, but because differing subsets of parameters will be combined additively, more than $n - 1$ distinct values in a matrix fitted to $\mathbf{P}$ is possible. The resulting fitted matrix will be SAR, so a consistent graphical representation can be provided, and ultrametrics will be included as a special case.

The M-file we propose to construct the collection of partitions respecting the given object order is called `partitionfnd_averages.m`, and uses dynamic programming to construct a set of partitions with from 1 to n ordered classes (see Hubert, Arabie, & Meulman, 2001). The criterion minimized is the maximum over clusters of the average proximities within subsets. In the verbatim listing below, we note that the collection of partitions constructed is hierarchical and actually produces the same order-constrained classification as that discussed in Section 3.2. This will not necessarily (or even usually) be the case for other data sets that we might consider.

```
>> help partitionfnd_averages.m

   PARTITIONFND_AVERAGES uses dynamic programming to
   construct a linearly constrained cluster analysis that
   consists of a collection of partitions with from 1 to
   n ordered classes.

   syntax: [membership,objectives] = partitionfnd_averages(prox)

   PROX is the input proximity matrix (with a zero main diagonal
   and a dissimilarity interpretation);
   MEMBERSHIP is the n x n matrix indicating cluster membership,
   where rows correspond to the number of ordered clusters,
   and the columns are in the identity permutation input order
   used for PROX.
   OBJECTIVES is the vector of merit values minimized in the
   construction of the ordered partitions, each defined by the
   maximum over clusters of the average proximities within subsets.

>> [membership,objectives] = partitionfnd_averages(supreme_agree)

membership =

     1    1    1    1    1    1    1    1    1
     2    2    2    2    1    1    1    1    1
     3    3    3    3    2    2    2    1    1
     4    3    3    3    2    2    2    1    1
     5    4    4    4    3    2    2    1    1
```

```
       6    5    4    4    3    2    2    1    1
       7    6    5    5    4    3    2    1    1
       8    7    6    5    4    3    2    1    1
       9    8    7    6    5    4    3    2    1


objectives =

     0.5044
     0.3470
     0.3133
     0.2833
     0.2633
     0.2300
     0.2200
     0.2100
          0

>> member = membership(2:9,:)

member =

       2    2    2    2    1    1    1    1    1
       3    3    3    3    2    2    2    1    1
       4    3    3    3    2    2    2    1    1
       5    4    4    4    3    2    2    1    1
       6    5    4    4    3    2    2    1    1
       7    6    5    5    4    3    2    1    1
       8    7    6    5    4    3    2    1    1
       9    8    7    6    5    4    3    2    1

>> [fitted,vaf,weights,end_condition] = partitionfit(supreme_agree,member)

fitted =

        0    0.3633  0.3633  0.3633  0.6405  0.6405  0.6405  0.6405  0.6405
   0.3633       0    0.2850  0.2850  0.6405  0.6405  0.6405  0.6405  0.6405
   0.3633  0.2850       0    0.2200  0.6405  0.6405  0.6405  0.6405  0.6405
   0.3633  0.2850  0.2200       0    0.6405  0.6405  0.6405  0.6405  0.6405
   0.6405  0.6405  0.6405  0.6405       0    0.3100  0.3100  0.4017  0.4017
   0.6405  0.6405  0.6405  0.6405  0.3100       0    0.2300  0.4017  0.4017
   0.6405  0.6405  0.6405  0.6405  0.3100  0.2300       0    0.4017  0.4017
   0.6405  0.6405  0.6405  0.6405  0.4017  0.4017  0.4017       0    0.2100
   0.6405  0.6405  0.6405  0.6405  0.4017  0.4017  0.4017  0.2100       0


vaf =

     0.7369


weights =

     0.2388
     0.0383
     0.0533
     0.0250
     0.0550
     0.0100
     0.0100
     0.2100


end_condition =

      0
```

3.4 EXTENSIONS TO ADDITIVE TREES: INCORPORATING CENTROID METRICS

A currently popular alternative to the use of a simple ultrametric in classification, which might be considered an extension of the notion of an ultrametric, is that of an additive tree metric (again, comprehensive discussions can be found throughout all of Barthélemy & Guénoche, 1991). Generalizing the earlier characterization of an ultrametric, an $n \times n$ matrix $D = \{d_{ij}\}$ can be called an *additive tree metric* (*matrix*) if the ultrametric inequality condition (C) is replaced by $d_{ij} + d_{kl} \leq \max\{d_{ik} + d_{jl}, d_{il} + d_{jk}\}$ for $1 \leq i, j, k, l \leq n$ (the additive tree metric inequality). Or equivalently, for any object quadruple O_i, O_j, O_k, and O_l, the largest two values among the sums $d_{ij} + d_{kl}$, $d_{ik} + d_{jl}$, and $d_{il} + d_{jk}$ are equal. Any additive tree metric matrix D can be represented (in many ways) as a sum of two matrices, say $U = \{u_{ij}\}$ and $C = \{c_{ij}\}$, where U is an ultrametric matrix, and $c_{ij} = g_i + g_j$ for $1 \leq i \neq j \leq n$ and $c_{ii} = 0$ for $1 \leq i \leq n$, based on some set of values $g_1, \ldots, g_n$. The multiplicity of such possible decompositions results from the choice of where to place the root in the type of graphical representation we give in Figure 3.4.

To eventually construct the type of graphical additive tree representation of Figure 3.4, the process followed is first to graph the dendrogram induced by U, where (as for any ultrametric) the chosen root is equidistant from all terminal nodes. The branches connecting the terminal nodes are then lengthened or shortened depending on the signs and absolute magnitudes of $g_1, \ldots, g_n$. If one were willing to consider the (arbitrary) inclusion of a sufficiently large additive constant to the entries in D, the values of $g_1, \ldots, g_n$ could be assumed nonnegative. In this case, the matrix C would represent what is called a *centroid metric*, and although a nicety (particularly for some of the graphical representations we give later in avoiding the issue of presenting negative branch lengths), such a restriction is not absolutely necessary for the extensions we pursue. In any case, the number of parameters in the extended sense alluded to earlier that an additive tree metric requires could be equated to the maximum number of branch lengths that a representation such as Figure 3.4 might necessitate (i.e., n branches attached to the terminal nodes, and $n - 3$ to the internal nodes only, for a total of $2n - 3$).

One of the difficulties with working with additive trees and displaying them graphically is to find some sensible spot to site a root for the tree. Depending on where the root is placed, a differing decomposition of D into an ultrametric and a centroid metric is implied. The ultrametric components induced by the choice of root can differ widely, with major substantive differences in the branching patterns of the hierarchical clustering. The two M-files discussed below, `cent_ultrafnd_confit.m` and `cent_ultrafnd_confnd.m`, both identify best-fitting additive trees to a given proximity matrix but where the terminal nodes of (an) ultrametric portion of the fitted matrix are then ordered according to a constraining order (`conperm`) that is either input (in `cent_ultrafnd_confit.m`) or is identified as a good one to use (in `cent_ultrafnd_confnd.m`) and then given as an output vector. In both cases, a centroid metric is first fit to the input proximity matrix; the residual matrix is carried

Figure 3.4   Graph-theoretic representation for the ordered-constrained additive tree described in the text (having a VAF of 98.41%).

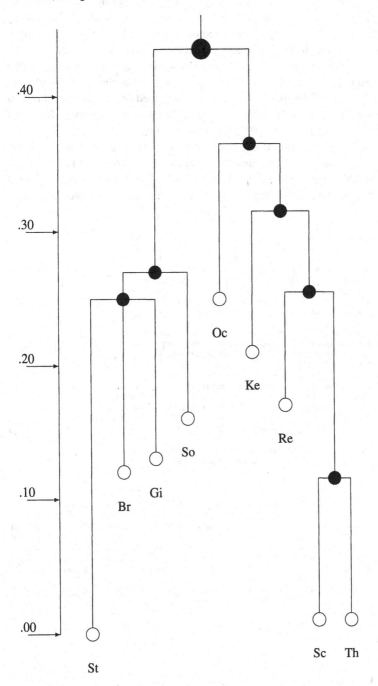

over to the order-constrained ultrametric identifying routines (`ultrafnd_confit.m` or `ultrafnd_confnd.m`), and thus, the root is chosen naturally for the ultrametric component. The entire process then iterates with a new centroid metric estimation, an order-constrained ultrametric reestimation, and so on, until convergence is achieved for the VAF values.

We illustrate below what occurs for our `supreme_agree` data and the imposition of the identity permutation (1 : 9) for the terminal nodes of the ultrametric. The relevant outputs are the ultrametric component in `targtwo` and the lengths for the centroid metric in `lengthsone`. To graph the additive tree, we first add .60 to the entries in `targtwo` to make them all positive and graph this ultrametric as in Figure 3.5. Then (1/2)(.60) = .30 is subtracted from each term in `lengthsone`; the branches attached to the terminal nodes of the ultrametric are then stretched or shrunk accordingly to produce Figure 3.4. [These stretching/shrinking factors are as follows: St: (.07); Br: (−.05); Gi: (−.06); So: (−.09); Oc: (−.18); Ke: (−.14); Re: (−.10); Sc: (.06); Th: (.06)]. We note that if `cent_ultrafnd_confnd.m` were used to find a good constraining order for the ultrametric component, the VAF could be increased slightly (to 98.56%) when the conperm of [3 1 4 2 5 6 7 9 8] is used. No real substantive interpretative difference, however, is apparent from the structure given in Figure 3.4.

```
>> help cent_ultrafnd_confit.m

    CENT_ULTRAFND_CONFIT finds and fits an additive tree by first fitting
    a centroid metric and secondly an ultrametric to the residual
    matrix where the latter is constrained by a given object order.

    syntax: [find,vaf,outperm,targone,targtwo,lengthsone] = ...
        cent_ultrafnd_confit(prox,inperm,conperm)

    PROX is the input proximity matrix (with a zero main diagonal
    and a dissimilarity interpretation); CONPERM is the given
    input constraining order (permutation) which is also given
    as the output vector OUTPERM;
    INPERM is a permutation that determines the order in which the
    inequality constraints are considered in identifying the ultrametric;
    FIND is the found least-squares matrix (with variance-accounted-for
    of VAF) to PROX satisfying the additive tree constraints. TARGTWO is
    the ultrametric component of the decomposition; TARGONE is the centroid
    metric component defined by the lengths in LENGTHSONE.

>> [find,vaf,outperm,targone,targtwo,lengthsone]
    = cent_ultrafnd_confit(supreme_agree,randperm(9),1:9)

find =
```

0	0.3800	0.3707	0.3793	0.6307	0.6643	0.7067	0.8634	0.8649
0.3800	0	0.2493	0.2579	0.5093	0.5429	0.5852	0.7420	0.7434
0.3707	0.2493	0	0.2428	0.4941	0.5278	0.5701	0.7269	0.7283
0.3793	0.2579	0.2428	0	0.4667	0.5003	0.5427	0.6994	0.7009
0.6307	0.5093	0.4941	0.4667	0	0.2745	0.3168	0.4736	0.4750
0.6643	0.5429	0.5278	0.5003	0.2745	0	0.2483	0.4051	0.4065
0.7067	0.5852	0.5701	0.5427	0.3168	0.2483	0	0.3293	0.3307
0.8634	0.7420	0.7269	0.6994	0.4736	0.4051	0.3293	0	0.2100
0.8649	0.7434	0.7283	0.7009	0.4750	0.4065	0.3307	0.2100	0

```
vaf =
```

Figure 3.5   Dendrogram representation for the ordered-constrained ultrametric component of the tree represented in Figure 3.4

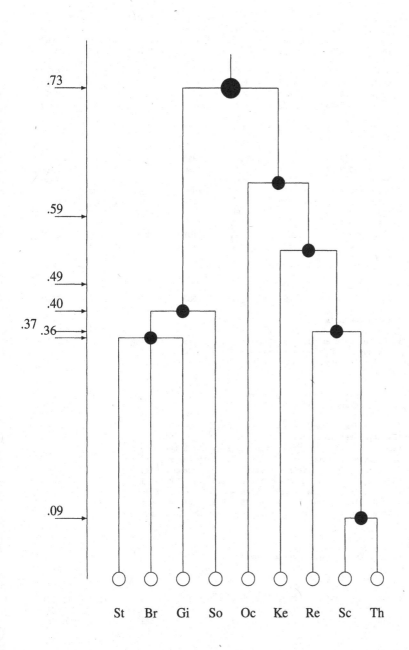

```
    0.9841


outperm =

    1    2    3    4    5    6    7    8    9


targone =

        0    0.6246    0.6094    0.5820    0.4977    0.5313    0.5737    0.7304    0.7319
   0.6246         0    0.4880    0.4606    0.3763    0.4099    0.4522    0.6090    0.6104
   0.6094    0.4880         0    0.4454    0.3611    0.3948    0.4371    0.5939    0.5953
   0.5820    0.4606    0.4454         0    0.3337    0.3673    0.4097    0.5664    0.5679
   0.4977    0.3763    0.3611    0.3337         0    0.2830    0.3253    0.4821    0.4836
   0.5313    0.4099    0.3948    0.3673    0.2830         0    0.3590    0.5158    0.5172
   0.5737    0.4522    0.4371    0.4097    0.3253    0.3590         0    0.5581    0.5595
   0.7304    0.6090    0.5939    0.5664    0.4821    0.5158    0.5581         0    0.7163
   0.7319    0.6104    0.5953    0.5679    0.4836    0.5172    0.5595    0.7163         0


targtwo =

        0   -0.2446   -0.2387   -0.2027    0.1330    0.1330    0.1330    0.1330    0.1330
  -0.2446         0   -0.2387   -0.2027    0.1330    0.1330    0.1330    0.1330    0.1330
  -0.2387   -0.2387         0   -0.2027    0.1330    0.1330    0.1330    0.1330    0.1330
  -0.2027   -0.2027   -0.2027         0    0.1330    0.1330    0.1330    0.1330    0.1330
   0.1330    0.1330    0.1330    0.1330         0   -0.0085   -0.0085   -0.0085   -0.0085
   0.1330    0.1330    0.1330    0.1330   -0.0085         0   -0.1107   -0.1107   -0.1107
   0.1330    0.1330    0.1330    0.1330   -0.0085   -0.1107         0   -0.2288   -0.2288
   0.1330    0.1330    0.1330    0.1330   -0.0085   -0.1107   -0.2288         0   -0.5063
   0.1330    0.1330    0.1330    0.1330   -0.0085   -0.1107   -0.2288   -0.5063         0


lengthsone =

    0.3730    0.2516    0.2364    0.2090    0.1247    0.1583    0.2007    0.3574    0.3589

>> help cent_ultrafnd_confnd.m

   CENT_ULTRAFND_CONFND finds and fits an additive tree by first fitting
   a centroid metric and secondly an ultrametric to the residual
   matrix where the latter is displayed by a constraining object order that
   is also identified in the process.

   syntax: [find,vaf,outperm,targone,targtwo,lengthsone] = ...
       cent_ultrafnd_confnd(prox,inperm)

   PROX is the input proximity matrix (with a zero main diagonal
   and a dissimilarity interpretation);
   INPERM is a permutation that determines the order in which the
   inequality constraints are considered in identifying the ultrametric;
   FIND is the found least-squares matrix (with variance-accounted-for
   of VAF) to PROX satisfying the additive tree constraints. TARGTWO is
   the ultrametric component of the decomposition; TARGONE is the centroid
   metric component defined by the lengths in LENGTHSONE; OUTPERM is the
   identified constraining object order used to display the ultrametric
   component.

>> [find,vaf,outperm,targone,targtwo,lengthsone]
   = cent_ultrafnd_confnd(supreme_agree,randperm(9))

find =

        0    0.3400    0.2271    0.2794    0.4974    0.5310    0.5734    0.7316    0.7301
   0.3400         0    0.3629    0.4151    0.6331    0.6667    0.7091    0.8673    0.8659
   0.2271    0.3629         0    0.2556    0.4736    0.5072    0.5495    0.7078    0.7063
```

0.2794	0.4151	0.2556	0	0.4967	0.5303	0.5727	0.7309	0.7294
0.4974	0.6331	0.4736	0.4967	0	0.2745	0.3168	0.4750	0.4736
0.5310	0.6667	0.5072	0.5303	0.2745	0	0.2483	0.4065	0.4051
0.5734	0.7091	0.5495	0.5727	0.3168	0.2483	0	0.3307	0.3293
0.7316	0.8673	0.7078	0.7309	0.4750	0.4065	0.3307	0	0.2100
0.7301	0.8659	0.7063	0.7294	0.4736	0.4051	0.3293	0.2100	0

vaf =

0.9856

outperm =

| 3 | 1 | 4 | 2 | 5 | 6 | 7 | 9 | 8 |

targone =

0	0.6151	0.4556	0.4787	0.3644	0.3980	0.4404	0.5986	0.5971
0.6151	0	0.5913	0.6144	0.5001	0.5337	0.5761	0.7343	0.7329
0.4556	0.5913	0	0.4549	0.3406	0.3742	0.4165	0.5748	0.5733
0.4787	0.6144	0.4549	0	0.3637	0.3973	0.4397	0.5979	0.5964
0.3644	0.5001	0.3406	0.3637	0	0.2830	0.3253	0.4836	0.4821
0.3980	0.5337	0.3742	0.3973	0.2830	0	0.3590	0.5172	0.5158
0.4404	0.5761	0.4165	0.4397	0.3253	0.3590	0	0.5595	0.5581
0.5986	0.7343	0.5748	0.5979	0.4836	0.5172	0.5595	0	0.7163
0.5971	0.7329	0.5733	0.5964	0.4821	0.5158	0.5581	0.7163	0

targtwo =

0	-0.2751	-0.2284	-0.1993	0.1330	0.1330	0.1330	0.1330	0.1330
-0.2751	0	-0.2284	-0.1993	0.1330	0.1330	0.1330	0.1330	0.1330
-0.2284	-0.2284	0	-0.1993	0.1330	0.1330	0.1330	0.1330	0.1330
-0.1993	-0.1993	-0.1993	0	0.1330	0.1330	0.1330	0.1330	0.1330
0.1330	0.1330	0.1330	0.1330	0	-0.0085	-0.0085	-0.0085	-0.0085
0.1330	0.1330	0.1330	0.1330	-0.0085	0	-0.1107	-0.1107	-0.1107
0.1330	0.1330	0.1330	0.1330	-0.0085	-0.1107	0	-0.2288	-0.2288
0.1330	0.1330	0.1330	0.1330	-0.0085	-0.1107	-0.2288	0	-0.5063
0.1330	0.1330	0.1330	0.1330	-0.0085	-0.1107	-0.2288	-0.5063	0

lengthsone =

| 0.2397 | 0.3754 | 0.2159 | 0.2390 | 0.1247 | 0.1583 | 0.2007 | 0.3589 | 0.3574 |

Lawrence Hubert
University of Illinois, Champaign-Urbana

Hans-Friedrich Köhn and Douglas Steinley
University of Missouri, Columbia

REFERENCES

Barthélemy, J.-P., & Guénoche, A. (1991). *Trees and proximity representations*. Chichester, UK: Wiley.

Bar-Joseph, Z., Gifford, D., & Jaakkola, T. S. (2001). Fast optimal leaf ordering for hierarchical clustering. *Bioinformatics*, *17*, S22–S29.

Caraux, G., & Pinloche, S. (2005). PermutMatrix: a graphical environment to arrange gene expression profiles in optimal linear order. *Bioinformatics*, *21*, 1280–1281.

Degerman, R. (1982). Optimal binary trees constructed through an application of Kendall's tau. *Psychometrika*, *47*, 523–527.

Dykstra, R. L. (1983). An algorithm for restricted least squares regression. *Journal of the American Statistical Association*, *78*, 837–842.

Gruvaeus, G., & Wainer, H. (1972). Two additions to hierarchical cluster analysis. *British Journal of Mathematical and Statistical Psychology*, *25*, 200–206.

Hubert, L., Arabie, P., & Meulman, J. (1998). Graph-theoretic representations for proximity matrices through strongly-anti-Robinson or circular strongly-anti-Robinson matrices. *Psychometrika*, *63*, 341–358.

Hubert, L., Arabie, P., & Meulman, J. (2001). *Combinatorial data analysis: Optimization by dynamic programming*. SIAM Monographs on Discrete Mathematics and Applications. Philadelphia: Society for Industrial and Applied Mathematics.

Hubert, L., Arabie, P., & Meulman, J. (2006). *The structural representation of proximity matrices with MATLAB*. ASA-SIAM Series on Statistics and Applied Probability. Philadelphia: Society for Industrial and Applied Mathematics.

Parmigiani, G., Garrett, E. S., Irizarry, R. A., & Zeger, S. L. (2003). The analysis of gene expression data: An overview of methods and software. In G. Parmigiani, E. S. Garrett, R. A. Irizarry, and S. L. Zeger (Eds.), *The analysis of gene expression data: methods and software* (pp. 1–45). New York: Springer-Verlag.

Wollan, P. C., & Dykstra, R. L. (1987). Minimizing linear inequality constrained Mahalanobis distances. *Applied Statistics*, *36*, 234–240.

CHAPTER 4

MULTIOBJECTIVE MULTIDIMENSIONAL (CITY-BLOCK) SCALING

4.1 INTRODUCTION

Combinatorial data analysis (CDA) encompasses a wide class of models and methods that can be applied to a number of quantitative problems in psychology, as well as other disciplines. For example, Arabie and Hubert (1992) outlined important CDA problems related to seriation, cluster analysis, additive trees, and network models. Their review captured both the various classes of combinatorial optimization problems in each of these areas, as well as available solution procedures for such problems. In a recent monograph, Hubert, Arabie, and Meulman (2001) provided an updated review of hierarchical clustering, partitioning, and seriation, as well as a general dynamic programming paradigm that can be used to obtain optimal solutions for problems of modest size. Recent developments in CDA have also emphasized the need for multiobjective programming methods that enable the quantitative analyst to make trade-offs among competing objective criteria. Delattre and Hansen (1980) presented an algorithm for a bicriterion partitioning problem that focused on the identification of clusters that were both homogeneous and well separated. Ferligoj and Batagelj (1992) described a number of approaches for multiobjective programming in cluster analysis, emphasizing a direct clustering algorithm for multiobjective hierarchical clustering.

More recently, Krieger and Green (1996), DeSarbo and Grisaffe (1998), Brusco, Cradit, and Stahl (2002), and Brusco, Cradit, and Tashchian (2003) have presented multiobjective partitioning models that seek to identify homogeneous clusters, while maintaining sufficient explanation of one or more exogenous dependent variables in each of those clusters.

Brusco and Stahl (2001) presented a multiobjective dynamic programming procedure for seriation that can be used to identify a permutation of objects that provides a good structural fit to a single proximity matrix. The objective function of their model was to optimize a weighted function of two or more structural indices. Within the context of seriation of a set of M proximity matrices, $\mathcal{A} = \{A_1, A_2, \ldots, A_M\}$, available for the same stimulus set, Brusco (2002) observed that an optimal permutation of objects for any particular matrix in $\mathcal{A}$, or a matrix based on pooling the matrices in $\mathcal{A}$, can provide a poor representation for one or more of the other matrices in $\mathcal{A}$. He subsequently proposed a multiobjective programming model for finding a permutation that provides a good fit for each of the matrices in $\mathcal{A}$. Using the multiobjective programming model and a dynamic programming-based solution algorithm, Brusco demonstrated the possibility of finding a single permutation that fit each of the matrices in A extremely well, even when antagonistic relationships existed among some of the matrices in $\mathcal{A}$. Motivated by the recent success for multiobjective clustering and seriation, our current paper focuses on a natural extension of the multiobjective programming paradigm to another important CDA application in multidimensional scaling (MDS). Although many MDS problems are particularly amenable to continuous optimization techniques, the inherent combinatorial structure of city-block MDS is well recognized (Brusco, 2001; Carroll & Arabie, 1998; Heiser, 1989; Hubert & Arabie, 1988; Hubert, Arabie, & Hesson-McInnis, 1992). The extension of the multiobjective CDA paradigm to city-block MDS presents a formidable challenge because we must obtain object permutations on multiple dimensions, as well as coordinates for each object on each dimension. Dynamic programming-based optimal solution procedures for multiobjective seriation are not feasible for city-block MDS because the inherent recursive structure is no longer present, and heuristic methods are therefore necessary.

Despite the computational challenges, extension of the multiobjective programming paradigm to city-block MDS is particularly worthwhile because of the well-recognized problems of pooling across multiple proximity matrices in MDS analyses (Ashby, Maddox, & Lee, 1994; Furnas, 1989; Lee & Pope, 2003; Siegler, 1987). Although the multiobjective programming approach described herein does not provide a definitive remedy for the problem of pooling across many subjects, our proposed method will enable quantitative analysts to apply a systematic procedure for obtaining good structural fits for each of several proximity matrices. As noted by Davison (1983, pp. 121-122), there are frequently situations where researchers collect data for differing experimental settings, treatment conditions, or other collection occasions. The proximity matrices obtained from the multiple sources are subsequently pooled for further analysis. Examples from the literature include pooling across different voices in auditory recognition tasks (Morgan, Chambers, & Morton, 1973), and pooling across active and passive touch in tactile experiments (Vega-Bemudez, Johnson, & Hsiao, 1991).

The multiobjective programming models for city-block MDS that are presented in this chapter provide sufficient flexibility to handle either a metric loss function or a stress measure (Kruskal, 1964a,b) for nonmetric relaxations. Solutions to the multiobjective problems are provided by a heuristic procedure that is based on previously published combinatorial methods for city-block MDS (Brusco, 2001; Hubert et al., 1992). To demonstrate their efficacy, the multiobjective models are applied to two published data sets from the psychological literature. A brief review of city-block MDS is reported in Section 4.2. In Section 4.3 we present the multiobjective city-block MDS model and in Section 4.4 provide a description of the solution algorithm for the multiobjective problem. Demonstrations of the multiobjective programming model are presented in Section 4.5, and the chapter concludes with a brief summary in Section 4.6.

4.2 CITY-BLOCK MDS

When the dimensions of psychological stimuli are separable, a common observation with MDS models is that city-block distances provide the preferred representation of stimulus space (Arabie, 1991; Attneave, 1950; Carroll & Arabie, 1980, 1998; Garner, 1974; Lee, 2001; MacKay, 2001; Myung & Shepard, 1996; Nosofsky, 1992; Shepard, 1964, 1987; Shepard & Arabie, 1979; Suppes, Krantz, Luce, & Tversky, 1989). Research on the perception of objects is one particular area that provides strong support for the tenability of the city-block metric. For example, Borg and Leutner (1983) and Hubert et al. (1992) successfully fitted two-dimensional city-block models to dissimilarity data related to the perception of the similarity of rectangles. Similarly, Shepard and Cermak (1973) found city-block properties in judgments regarding toroidal forms, and Dunn (1983) observed that the dimensions of parallelogram size and tilt were combined on the basis of a city-block metric. More recently, results reported by Kruschke (1993) and Lee (2001) indicated the appropriateness of a city-block metric for data related to categorization tasks involving rectangles. Wuerger, Maloney, and Krauskopf (1995) found evidence that subject judgments regarding the proximity of colored lights were consistent with the city-block metric. The relevance of city-block MDS even extends to animal research. For example, Ronacher (1992) observed that bees used the city-block (as opposed to Euclidean) metric in their discrimination of ring patterns.

As noted by Arabie (1991, p. 572), even previous studies that have not revealed support for the city-block metric (e.g., Hoben, 1968; Melara, 1989; Monahan & Lockhead, 1977; Tversky & Gati, 1982) must be viewed cautiously because of an algorithmic issue. One of the well-documented problems of city-block MDS is the potential for relatively poor local optima, especially when traditional gradient-based methods are used (Arabie, 1973, 1991; Grau & Nelson, 1988; Groenen & Heiser, 1996; Hubert & Arabie, 1988; Hubert et al., 1992; Shepard, 1974). When employed to minimize loss functions (or Stress indices in the nonmetric case) in city-block

MDS, gradient-based methods exhibit convergence problems and often fail to move very far from the starting configuration. As a result, the solutions provided by gradient-based methods frequently yield loss functions (or Stress values) that are much larger than those corresponding to global optima. Hubert and Arabie (1988) and Hubert et al. (1992) provide particularly good discussions and demonstrations of the reason gradient-based methods are apt to fail in the two-dimensional city-block context.

The well-recognized problem of poor local minima has stimulated the development of a number of alternative solution procedures for city-block MDS, thus contributing to the effort to overcome the limitations of gradient-based methods and corresponding inferior local optima. In addition to combinatorial methods (Brusco, 2001; Heiser, 1989; Hubert et al., 1992), these procedures include majorization (Heiser, 1989), the tunneling method (Groenen & Heiser, 1996), and distance smoothing heuristics (Groenen, Heiser, & Meulman, 1998, 1999). Although none of these methods guarantee the identification of a globally optimal solution, simulation experiments do seem to suggest that they are at least capable of finding very good locally optimal solutions.

4.3 MULTIOBJECTIVE CITY-BLOCK MDS

4.3.1 The Metric Multiobjective City-Block MDS Model

We initially present a multiobjective programming model for multidimensional scaling in the city-block metric assuming a least-squares loss function (Brusco, 2001; Hubert et al., 1992). The extension of the model for nonmetric MDS, which assumes the optimization of a weighted stress measure, is straightforward and is described in Section 4.3.2.

We specify $S = \{1, 2, \ldots, n\}$ as the set of indices for a collection of n objects, and $\mathcal{A} = \{A_1, A_2, \ldots, A_M\}$ is more specifically defined as a collection of M symmetric $(n \times n)$ dissimilarity matrices with nonnegative off-diagonal elements $a_{ijm} = a_{jim}$ representing the dissimilarity between objects i and j for matrix m. We further denote Ψ as the set of all permutations of the n object indices, and $\psi_d = \{\psi_d(1), \psi_d(2), \ldots, \psi_d(n)\}$ as the permutation of the objects on dimension d for $d = 1, \ldots, D$. Hereafter, we refer to $\psi = \{\psi_1, \psi_2, \ldots, \psi_D\}$ as a dimension permutation set, containing the permutations for each dimension. The values of ϕ_{dj} are based on a one-to-one function with ψ_d such that $\phi_{dj} = k \ni \psi_d(k) = j$; thus, ϕ_{dj} is the sequence position of object j on dimension d. We also establish $\alpha_{dij} = \min(\phi_{di}, \phi_{dj})$ and $\beta_{dij} = (\max(\phi_{di}, \phi_{dj}) - 1)$ for $d = 1, \ldots, D$, $i = 1, \ldots, n-1$, and $j = i+1, \ldots, n$. The spacing (or distance) between objects in positions k and $k+1$ on dimension d is denoted as s_{dkm} for $k = 1, \ldots, n-1$, $d = 1, \ldots, D$, and $m = 1, \ldots, M$. Finally, we define user-specified parameters w_m and MU_m, which represent objective function weights and upper bounds on the least-squares loss function for matrix $m (m = 1, \ldots, M)$, respectively. Consistent with previous multiobjective programming implementations (Brusco, 2002; Brusco & Stahl, 2001), we select a set of positive weights that sum to 1. The least-squares loss functions for each of the M matrices can be considered as functions of the given dimension permutations as

$$f_m(\psi) = \sum_{i=1}^{n-1} \sum_{j=i+1}^{n} \left[a_{ijm} - \sum_{d=1}^{D} \sum_{k=\alpha_{dij}}^{\beta_{dij}} s_{dkm} \right]^2 \quad \text{for} \quad m = 1, \ldots, M$$

(4.1)

For each matrix m, $f_m(\psi)$ represents the sum-of-squared differences between the dissimilarity measures and the city-block distances between objects in multidimensional space. The distance for any pair of objects i and j on a particular dimension d is computed as the sum of the spacings for each pair of adjacent objects between the objects i and j. The city-block distance between i and j is then computed as the sum of the D dimension distances. For any fixed set of dimension permutations, the optimal spacings that minimize $f_m(\psi)$ can be obtained using a nonnegative least-squares algorithm (Hubert et al., 1992; Lawson & Hanson, 1974, pp. 304–309). The search for a global minimum solution for $f_m(\psi)$ can be conducted by evaluating different dimension permutations. Unfortunately, this process is inherently combinatorial in nature, and we must therefore rely on heuristic solution methods. Our situation is further complicated by the fact that it is necessary to find a single-dimension permutation set (ψ) that provides a good structural fit for each matrix $m = 1, \ldots, M$. For this reason, we consider the following metric multiobjective city-block MDS optimization problem (MCB1):

$$\min_{(\psi = \{\psi_1, \psi_2, \ldots, \psi_D \in \Psi\})} \sum_{m=1}^{M} w_m f_m(\psi)$$

(4.2)

$$\text{subject to:} \quad \sum_{m=1}^{M} w_m = 1$$

(4.3)

$$w_m > 0 \qquad \text{for} \quad m = 1, \ldots, M \quad (4.4)$$

$$f_m(\psi) \le MU_m \quad \text{for} \quad m = 1, \ldots, M \quad (4.5)$$

The objective function (4.2) of MCB1 is to minimize a weighted function of least-squares loss values across the M dissimilarity matrices. This weighted function is minimized subject to constraints (4.3) and (4.4), which guarantee the selection of positive weights that sum to 1. Constraint (4.5) allows the specification of a raw upper bound for one or more loss functions. In many implementations, the typical assumption is that $MU_m = \infty$ for all $m = 1, \ldots, M$. If it were possible to obtain optimal solutions for MCB1 for a given set of weights, the resulting solution would be a nondominated (or efficient) solution for the multiobjective problem. A solution corresponding to the dimension permutation set ψ is dominated by a solution corresponding to another dimension permutation set ψ' if $f(\psi') \le f(\psi)$ for $m = 1, \ldots, M$, and $f(\psi') < f(\psi)$ for at least one m. Unfortunately, obtaining optimal solutions to MCB1 is not practical, and the resulting heuristic solutions for different sets of weights are only providing estimates of the efficient frontier.

In some circumstances, one beneficial tactic is to modify the objective function of MCB1 with normalized loss functions. Specifically, using an effective city-block MDS heuristic procedure, a quantitative analyst could obtain near-optimal solutions for each of the single-objective problems associated with matrices $m = 1, \ldots, M$. Denoting the best-found loss function identified for matrix m as f_m^*, an alternative objective function is

$$\max_{(\psi = \{\psi_1, \psi_2, \ldots, \psi_D \in \Psi\})} \sum_{m=1}^{M} w_m \left(f_m^* / f_m(\psi) \right) \tag{4.6}$$

This normalized objective function, which is comparable to those used in recent multiobjective data analysis implementations (Brusco, 2002; Brusco & Stahl, 2001), expresses the loss function value for a matrix in relation to the best-found loss function for that matrix. One advantage of this representation is that, assuming the best-found solutions provide optimal index values, possible values of (4.6) range from zero to 1. For the rare situations where $f_m^* = 0$, we recommend eliminating the component for matrix m from the objective function and incorporating a constraint of type (4.5) with some appropriate MU_m. Hereafter, we refer to the normalized metric city-block MDS model as MCB2.

4.3.2 The Nonmetric Multiobjective City-Block MDS Model

Because of the prevalence of nonmetric MDS (Kruskal, 1964a,b; Shepard, 1962a,b) in the psychological literature, we deemed it necessary to allow for nonmetric relaxations of the proximity data when using the multiobjective paradigm. For nonmetric MDS, we can replace the least-squares loss function (4.1) with a stress function for matrix m. For example, using Kruskal's (1964a,b) stress measure (stress type 1), we have

$$g_m(\psi) = \sqrt{\frac{\sum_{i=1}^{n-1} \sum_{j=i+1}^{n} \left(\left(\sum_{d=1}^{D} \sum_{k=\alpha_{dij}}^{\beta_{dij}} s_{dkm} \right) - \hat{\alpha}_{ijm} \right)^2}{\sum_{i=1}^{n-1} \sum_{j=i+1}^{n} \left(\sum_{d=1}^{D} \sum_{k=\alpha_{dij}}^{\beta_{dij}} s_{dkm} \right)^2}} \tag{4.7}$$

for $m = 1, \ldots, M$, where the $\hat{\alpha}_{ijm}$ values are the transformed proximities (or disparities) that, using monotone regression, are obtained so as to preserve the same order as the original proximities. The nonmetric multiobjective city-block MDS optimization problem (NMCB1) is represented as

$$\min_{(\psi = \{\psi_1, \psi_2, \ldots, \psi_D \in \Psi\})} \sum_{m=1}^{M} w_m g_m(\psi) \tag{4.8}$$

$$\text{subject to:} \quad f_m(\psi) \le ML_m \quad \text{for } m = 1, \ldots, M \tag{4.9}$$

and (4.3) and (4.4), where ML_m is an upper bound on the stress for matrix m. Problem NMCB1 is also an extremely difficult combinatorial optimization problem and we must resort to heuristic procedures. As was the case for MCB1, we can modify the objective function to capture normalized objective function values as follows:

$$\max_{(\psi=\{\psi_1,\psi_2,...,\psi_D\in\Psi\})} \sum_{m=1}^{M} w_m \left(g_m^*/g_m(\psi)\right) \qquad (4.10)$$

where g_m^* represents the best stress value found by applying an appropriate heuristic to the single-objective problem posed by matrix m. We refer to the normalized nonmetric city-block model as NMCB2.

4.4 COMBINATORIAL HEURISTIC

Hubert et al. (1992) provided convincing arguments regarding the inherent combinatorial structure of city-block MDS, specifically observing that the problem can be decomposed into two interrelated stages: (a) finding an appropriate permutation for each dimension, and (b) solving for the optimal coordinates given a set of fixed dimension permutations. The method we employ for multiobjective city-block MDS is a direct extension of Hubert et al.'s (1992) combinatorial heuristic paradigm, and consists of the following steps:

Step 1. Generate an initial permutation of objects, ψ_d, for each dimension $d = 1,\ldots,D$.

Step 2. For each dissimilarity matrix, use nonnegative least-squares to find the optimal spacing between each pair of objects on each dimension, s_{dkm}, so as to minimize the least-squares difference between the dissimilarities and distances. If a nonmetric relaxation is desired, go to step 5; otherwise, go to step 3.

Step 3. Compute the multiobjective loss function value (4.2) or the normalized index (4.6), and store this value and corresponding dimension permutations as the incumbent solution.

Step 4. For each dimension, evaluate the effect of all possible pairwise interchanges of objects by cycling through steps 2 and 3. Each time an interchange is identified such that the value of (4.2) or (4.6) is improved, accept the interchange and store the new solution and its corresponding objective function value as the incumbent solution. Continue this process until no pairwise interchange of objects on any dimension will improve the value of (4.2) or (4.6). The algorithm terminates when no pairwise interchange will further improve the objective function.

Step 5. For each dissimilarity matrix, use iterative monotone regression to find the disparities, $\hat{a}_{ijm}$, that minimize stress (4.7).

Step 6. Compute the multiobjective stress function value (4.8) or the normalized index (4.10), and store this value and corresponding dimension permutations as the incumbent solution.

Step 7. For each dimension, evaluate the effect of all possible pairwise interchanges of objects by cycling through steps 2, 5, and 6. Each time an interchange is identified such that the value of (4.8) or (4.10) is improved, accept the interchange and store the new solution and its corresponding objective function value as the incumbent solution. Continue this process until no pairwise interchange of objects on any dimension will improve the value of (4.8) or (4.10). The algorithm terminates when no pairwise interchange will further improve the objective function.

The heuristic algorithm begins in step 1 by obtaining a permutation of objects on each dimension. These initial permutations can be obtained by randomly selecting object positions for each dimension from a uniform distribution of the integers $[1, n]$ (without replacement). Alternatively, much better starting solutions can be obtained by using lattice-search heuristics for placement of the objects in continuous space (Brusco, 2001; Hubert & Busk, 1976); however, these heuristics substantially increase computational effort. In step 2, for each dissimilarity matrix, a nonnegative least-squares algorithm (Lawson & Hanson, 1974, pp. 304–309) was used to obtain the optimal spacing between each pair of adjacent objects on each dimension. These spacings are computed such that the squared differences between the city-block distances (which are an additive function of the distances) and the dissimilarity measures are minimized.

In the case of metric city-block MDS, the weighted least-squares loss function (4.2) or the normalized index (4.6) for the multiobjective problem is computed in step 3. Pairwise interchange of objects on each dimension are evaluated at step 4, and each interchange that improves (4.2) or (4.6) is accepted. The algorithm terminates when no pairwise interchange of objects on any dimension will further improve (4.2) or (4.6). Pairwise interchange can be augmented by other local search operations, such as object block reversals and object insertions (Hubert & Arabie, 1994; Hubert, Arabie, & Meulman, 1997). Our investigation of these additional operations confirmed the findings of Hubert et al. (1992), which suggested that they tended to increase computation time with little or no improvement in solution quality.

If a nonmetric relaxation is desired, then control passes from step 2 to step 5. At this point, iterative monotone regression is used to minimize Stress for each dissimilarity matrix. An up-and-down blocks algorithm (Kruskal, 1964b) is used to obtain the disparities, $\hat{a}_{ijm}$, and these disparities iteratively replace the dissimilarity measures until convergence of the stress measure (tolerance = .00001). The weighted stress function is obtained in step 6, and pairwise interchange is implemented in step 7. One important distinguishing factor of the nonmetric MDS algorithm used herein is that a relaxation is obtained after each pairwise interchange. This is slightly different from Hubert et al.'s (1992) original implementation, which employed monotone regression only after no pairwise interchanges could further improve the least-squares loss function. Our implementation is therefore much more computationally intensive than

Hubert et al.'s (1992) original procedure for single-objective problems, but we found that the additional calculation tended to be worthwhile for multiobjective problems.

Because the combinatorial heuristic is sensitive to the initial permutations, we use at least 20 replications of the heuristic in our experimental analyses. The choice regarding the number of replications is based on the number of objects, the number of dissimilarity matrices, the number of dimensions, and, most important, whether a nonmetric relaxation of the proximities is desired. Nonmetric implementations require considerably more CPU time than metric ones because of the need to use monotone regression when testing each pairwise interchange. Each replication uses a different randomly generated set of dimension permutations to initiate the heuristic. The combinatorial heuristic was written in Fortran and implemented on a 2.2-GHz Pentium II PC with 1 GB of RAM.

4.5 NUMERICAL EXAMPLES

4.5.1 Example 1

The first numerical example corresponds to auditory recognition data originally collected and analyzed by Morgan et al. (1973), and also studied by Hubert and Baker (1977) and Brusco (2002). The stimuli consist of spoken digits $1, 2, \ldots, 9$, and $M = 2$ dissimilarity matrices were derived from two auditory confusion matrices (9×9) among the nine digits (Morgan et al., 1973, p. 376). The two dissimilarity matrices were obtained from similar experiments, differentiated primarily by the gender of the person speaking the digits. It should also be noted that based on the results of cluster and nonmetric MDS analyses of the two matrices, Morgan et al. (1973, p. 379) "...felt justified in basing further analysis on the data obtained by pooling over voices to give a single confusion matrix for the recognition task." For this reason, we considered three dissimilarity matrices in our analyses: (a) *female*, the dissimilarity matrix associated with the female voice,; (b) *male*, the dissimilarity matrix associated with the male voice; and (c) *pooled*, a dissimilarity matrix obtained by pairwise averaging of the female and male dissimilarity matrices.

Nonmetric two-dimensional city-block solutions were obtained for each of the three matrices using the combinatorial heuristic developed by Hubert et al. (1992). For each matrix, the best local minimum (or best found) stress value was identified for at least 22 of the 100 replications. The best-found dimension permutations for each of the three matrices are shown in Table 4.1. Each of these dimension permutations were then used to obtain nonmetric two-dimensional fits for all three matrices. For example, the best-found dimension permutations for female were used to obtain nonmetric solutions for male and pooled. Applying this approach for each of the three best-found sets of dimension permutations resulted in a total of nine stress values that are shown in Table 4.2. The main diagonal of the stress value matrix shown in Table 4.2 represents the best-found stress value for the matrix, and the off-diagonal elements represent the stress value that results when the best-found dimension permutations corresponding to the data matrix associated with the row are used to fit the data matrix associated with the column.

Table 4.1 Best-Found Dimension Permutations for Each Dissimilarity Matrix and the Multiobjective Solution for Numerical Example 1

Best-found permutation for:		Permutation
Female recognition matrix	Dimension 1:	5-9-1-4-8-3-7-2-6
	Dimension 2:	4-7-5-1-6-2-9-3-8
Male recognition matrix	Dimension 1:	5-9-1-4-2-3-8-7-6
	Dimension 2:	4-8-6-5-1-9-7-3-2
Pooled recognition matrix	Dimension 1:	5-9-1-4-2-3-8-7-6
	Dimension 2:	4-8-5-6-1-7-9-3-2
Multiobjective solution using equal weights	Dimension 1:	5-9-1-4-3-8-7-2-6
	Dimension 2:	4-8-6-7-5-1-9-3-2

Table 4.2 Stress Values for Each Dissimilarity Matrix in Numerical Example 1 Obtained Using the Best-Found Dimension Permutations for Each Matrix

	Female	Male	Pooled
Using Female dim. permutations	.07305	.13704	.10869
Using Male dim. permutations	.10765	.05218	.06722
Using Pooled dim. permutations	.10541	.05705	.06322

The stress values in Table 4.2 reveal that reasonably good two-dimensional fits can be obtained for both the female and male dissimilarity matrices, as well as the pooled dissimilarity matrix. However, the off-diagonal elements suggest that the solutions for the pooled matrix more closely correspond to the solution for the male matrix. For example, a within-column examination of the submatrix of stress values corresponding to male and pooled reveals very little variation. This finding is attributable to the fact that the best-found dimension permutations for these two matrices are quite similar and, therefore, they yield rather similar solutions for each other. The best-found dimension permutations for the male and pooled matrices differ only by the interchange of adjacent objects 6 and 5 and adjacent objects 9 and 7 on the second dimension.

The best-found dimension permutations for the female dissimilarity matrix exhibit some marked differences from the corresponding permutations for the other two matrices. These differences manifest themselves in the form of somewhat larger stress values ($> .10$) when the dimension permutations for male and pooled are applied to the female matrix. Perhaps a clearer representation of the differences can be observed from Figure 4.1, which displays the two-dimensional solutions for female and male. Although there is clearly some similarity between the two solutions (e.g., the positions of objects 5, 9, 1, and 4), there are also some notable differences. Objects 2 and 3 are more separated from the remaining objects in the male solution, and object 8 is more centrally positioned among objects 4, 7, and 6 in the male solution.

The key finding in our analysis thus far is that the MDS solution for the pooled matrix closely resembles the MDS solution for the male matrix, but differs from the solution for the female matrix. Thus, pooling the data has incorporated information from the male matrix while mitigating the information from the female matrix. The multiobjective programming approach provides a means of possibly identifying dimension permutations that can yield good structural fits for both the female and male matrices. In the absence of any additional information, we began our multiobjective analysis using NMCB1 and weights of $w_1 = .5$ and $w_2 = .5$. This seemed logical because the best-found stress values for female and male were reasonably close, and we wanted to maintain good solutions for each. Methods for interactively searching for good weights are described by Brusco (2002) and Brusco and Stahl (2001).

Using weights of $w_1 = .5$ and $w_2 = .5$, one of the best multiobjective solutions obtained across 100 replications of the combinatorial heuristic provided stress values of .08434 and .06895 for the female and male dissimilarity matrices, respectively. The optimal permutation for the first dimension of the multiobjective solution was 5-9-1-4-3-8-7-2-6, whereas the second dimension permutation was 4-8-6-7-5-1-9-3-2. An interesting observation is that the permutation for the first dimension of the multiobjective solution is very close to the first-dimension of the female solution, as these permutations differ only by the interchange of adjacent objects 3 and 8. (The difference between the multiobjective solution and the first dimension 'male' solution is the relocation of object '2'.) The permutation for the second-dimension of the multiobjective solution is very close to the second dimension of the male solution, with the only difference being that object 7 moved up three positions in the multiobjective ordering. Several second-dimension differences exist between the multiobjective solution and the female solution. This multiobjective solution clearly illustrates the nature of the trade-offs between the female and male matrices.

Figure 4.2 presents the two-dimensional multiobjective city-block MDS solutions for the female and male dissimilarity matrices. The two maps in Figure 4.2 exhibit greater similarity than the two maps in Figure 4.1 because the dimension permutations are restricted to be the same. We observe that both of the maps in Figure 4.2 provide fairly strong separation of objects 2 and 3 from the remaining objects, as well as the centralized location of object 8 with objects 4, 7, and 6. Both of these properties are consistent with the male map from Figure 4.1. The influence from the female matrix in the multiobjective solution is observed by the placement of object 2 on dimension 1, as well as the placement of object 7 on dimension 2.

Figure 4.1 Two-dimensional nonmetric city-block MDS solution for male and female recognition matrices.

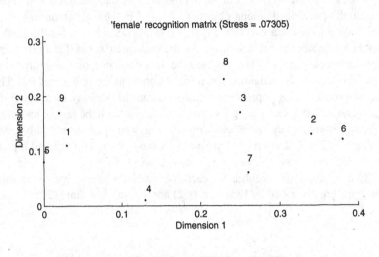

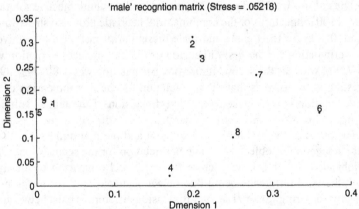

4.5.2 Example 2

For our second example we used data pertaining to visual identification tasks for textured materials originally collected by Cho, Yang, and Hallett (2000). The stimuli consist of $n = 20$ textured objects, and $M = 3$ dissimilarity matrices were derived from three confusion matrices (20×20) among the objects (Cho et al., 2000, pp. 750–751). Table 4.3 identifies the 20 objects using both their numeric code from previous studies (Cho et al., 2000; Rao & Lohse, 1996) as well as a letter representation used herein. The three confusion matrices correspond to the same identification task

Figure 4.2   Two-dimensional nonmetric city-block MDS solution from the multiobjective programming model for male and female recognition matrices.

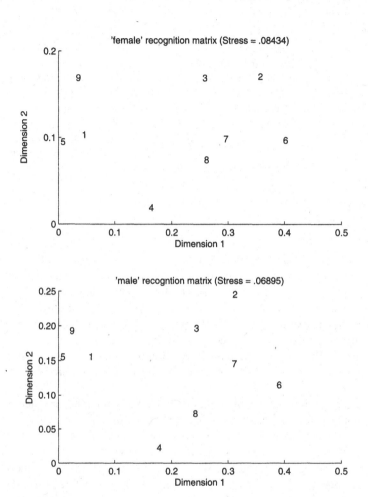

measured at three different distances (8.2, 15.5, and 22.9 m) between the subjects and the stimuli. The three confusion matrices were converted to symmetric dissimilarity matrices using the procedure described by Cho et al. (2000, p. 738). Specifically, each confusion matrix was added to its transpose to form a symmetric similarity matrix. The largest element in each similarity matrix (we ignored the main diagonal) was stored, and each of the dissimilarity matrices were formed by subtracting the similarity elements from the largest element associated with that matrix.

Table 4.3 Textures Used in Numerical Example 2

Label	Number	Description	Label	Number	Description
A	4	pressed cork	K	55	straw matting
B	9	grass lawn	L	66	plastic pellets
C	12	bark of tree	M	68	wood grain
D	16	herringbone weave	N	76	oriental glass fiber
E	20	French canvas	O	86	ceiling tile
F	22	reptile skin	P	87	sea fan
G	29	beach sand	Q	93	fur
H	34	netting	R	98	crushed rose quartz
I	37	water	S	109	handmade pepper
J	52	oriental straw cloth	T	112	plastic bubbles

Source: Cho et al. (2000).

Nonmetric MDS results reported by Cho et al. (2000) suggested the need for spatial representations of at least three dimensions. We therefore began our analyses by running 50 replications of Hubert et al.'s (1992) combinatorial heuristic to obtain single-objective, three-dimensional solutions for each of the three dissimilarity matrices. The minimum stress values across the 50 replications were .05647, .06653, and .04825 for the dissimilarity matrices corresponding to distances of 8.2, 15.5, and 22.9 m, respectively. These values compared favorably to graphically depicted stress results from Cho et al. (2000, p. 747) and supported the tenability of a three-dimensional solution.

We also used 50 replications of Hubert et al.'s (1992) combinatorial heuristic to produce a solution for the dissimilarity matrix obtained from pooling the dissimilarity matrices. The best solution identified across the 50 replications provided a Stress value of .09263 for the pooled dissimilarity matrix. When the dimension permutations corresponding to this solution were applied to each of the three individual matrices, the resulting stress values were .09421, .09945, and .07259, for the matrices corresponding to distances of 8.2, 15.5, and 22.9 m, respectively. Thus, for this particular set of dissimilarity matrices, the solution for the pooled matrix did not provide an especially poor fit for any of the matrices. Nevertheless, we implemented a three-dimensional tri-criterion version of the nonmetric multiobjective city-block MDS model in an effort to find a set of dimension permutations that yielded even lower stress values.

We applied 20 replications of the multiobjective combinatorial heuristic using both NMCB1 and NMCB2 and equal weights of $w_1 = w_2 = w_3 = 1/3$. For each of these replications, the stress function values for each of the three dissimilarity matrices are displayed in Table 4.4 and 4.5, respectively. These results suggest that the combinatorial heuristic tends to find good solutions on most replications, but the solutions for NMCB1 were typically superior. When using NMCB1 as the objective criterion, there were three replications that provided a set of stress values that dominated the solution obtained using the pooled matrix. One of these was a solution that provided stress values of .08786, .08528, and .06463, for the matrices corresponding to distances of 8.2, 15.5, and 22.9 m, respectively. Although a comprehensive analysis of visual recognition of textures is beyond the scope of this chapter, our multiobjective analyses did produce results consistent with those previously reported in the literature (Cho et al., 2000; Rao & Lohse, 1996).

Table 4.4   Multiobjective City-Block MDS Results for Texture Recognition Data Using 20 Replications of the Combinatorial Heuristic for NMCB1 Assuming Equal Weights

Repl.	8.2 m	15.5 m	22.9 m	Eq. (4.8)	Eq. (4.10)
1	.07354	.09996	.08335	.08561	.67071
2	.08472	.08374	.09964	.08936	.64836
3	.06877	.07951	.08841	.07889	.73448
4	.08158	.10804	.08635	.09198	.62219
5	.08464	.10085	.07989	.08845	.64354
6	.08268	.09219	.07201	.08229	.69150
7	.09797	.10174	.08519	.09496	.59884
8	.08946	.09310	.08459	.08904	.63868
9	.07651	.09461	.07569	.08226	.69285
10	.09586	.10307	.09517	.09802	.58046
11	.06675	.10975	.09375	.09007	.65555
12	.09338	.07494	.07696	.08175	.70642
13	.08786	.08528	.06463	.07925	.72307
14	.10391	.08083	.07734	.08735	.66340
15	.10114	.11012	.08607	.09910	.57430
16	.10667	.11119	.10961	.10915	.52259
17	.10011	.11932	.12464	.11468	.50287
18	.08788	.08933	.07770	.08496	.66937
19	.08094	.10443	.08834	.09123	.62692
20	.07682	.09820	.07026	.08175	.69970

Based on cluster analyses and MDS solutions obtained by Rao and Lohse (1996), dimension 1 might be described as *degree of coarseness*. According to the hierarchical clustering solution and attribute analyses reported by Rao and Lohse (p. 1656), the leftmost object on dimension 1 (object H, netting) is perceived as very fine, whereas the two rightmost objects (Q, fur and T, plastic bubbles) are perceived of as very course. The smallest coordinate for dimension 2 corresponds to object K, straw matting, which was in a group of objects perceived as highly regular in Rao and Lohse's study. Objects O, ceiling tile, and R, crushed rose quartz, have the largest coordinate values on dimension 2, and these objects tended to be characterized as highly irregular in Rao and Lohse's study. Thus, Dimension 2 might be interpreted as *degree of regularity*.

Table 4.5 Multiobjective City-Block MDS Results for Texture Recognition Data Using 20 Replications of the Combinatorial Heuristic for NMCB2 Assuming Equal Weights

Repl.	8.2 m	15.5 m	22.9 m	Eq. (4.8)	Eq. (4.10)
1	.08491	.14195	.15367	.12683	.48253
2	.07153	.14910	.14302	.12120	.52429
3	.09824	.09459	.11101	.10127	.57088
4	.08764	.11513	.11657	.10644	.54532
5	.08393	.10544	.10471	.09802	.58814
6	.05678	.13063	.11170	.09969	.64520
7	.10444	.09147	.07550	.09046	.63564
8	.09042	.08939	.07358	.08445	.67478
9	.09369	.14059	.14959	.12794	.46612
10	.10330	.07672	.07696	.08565	.68019
11	.06421	.13393	.12520	.10777	.58714
12	.12422	.08201	.06753	.09124	.66005
13	.10445	.07357	.06262	.08021	.73842
14	.09475	.09944	.09894	.09770	.58418
15	.10189	.09640	.10266	.10031	.57140
16	.09369	.10592	.10311	.10090	.56621
17	.06515	.08396	.09538	.08149	.72161
18	.07666	.10060	.08794	.08839	.64881
19	.07708	.10974	.09805	.09495	.61026
20	.10149	.12350	.11114	.11203	.50970

Finally, dimension 3 could be interpreted as *degree of directionality* because the objects with the smallest (I, water) and second-largest (O, ceiling tile) coordinates on this dimension were associated with the highest and lowest directionality groups in Rao and Lohse's study.

4.6 SUMMARY AND CONCLUSIONS

In this chapter, we have presented a multiobjective programming approach for multi-dimensional scaling in the city-block metric. The model is designed to find a permutation for each dimension that will enable a good structural fit to be obtained for each of a set of M dissimilarity matrices. A combinatorial heuristic for the multiobjective programming problem was presented, and demonstrations were provided for two sets of empirical data matrices from the literature. The results suggest that the multiobjective programming approach can find dimension permutations that yield good fits for multiple matrices. A related finding, which supports recent observations in the literature (Ashby, Maddox, & Lee 1994; Lee & Pope, 2003), is that pooling across multiple matrices can yield misleading results. Our findings suggest that solutions for pooled matrices might represent one matrix in the set very well, yet provide inferior solutions for others.

The biggest limitation of the multiobjective programming approach described in this chapter is its computational feasibility for large data sets. This is especially true for nonmetric analyses, which require monotone regression when testing pairwise interchanges. As noted by Hubert et al. (1992) and Brusco (2001), even single-objective combinatorial procedures can require considerable computational effort when $D > 2$. Although this problem is alleviated somewhat by significant improvements in processing speed, the need for more efficient optimization procedures persists. In this chapter, we provided a numerical demonstration for a three-dimensional ($D = 3$) nonmetric MDS scenario for a set of $n = 20$ objects measured on $M = 3$ dissimilarity matrices. Implementations for nonmetric MDS scenarios with values of D, M, and n that are larger than those in our demonstration would probably require considerable computation time. This is particularly true if the examination is of multiple sets of objective function weights. Fortunately, we were able to obtain good solutions using equal weights in both of our numerical examples; however, a more refined weighting scheme might be needed for other sets of data. For two to four matrices, finding a good set of weights does not present a daunting challenge. When attempting to find a set of weights that yields good solutions for more than four matrices, the search for such weights can be especially cumbersome.

One straightforward alternative for implementing the multiobjective procedure is to obtain a good solution for the pooled data matrix and then supply that solution as a starting point for the multiobjective programming model. We employed this strategy for the data in our second numerical example and were able to find a solution that was nearly as good as the best solution we found across 20 random starting solutions. Other methods for finding good initial permutations would undoubtedly be beneficial.

Although we have focused on city-block MDS, an analyst needs to recognize that the multiobjective programming paradigm presented herein is not necessarily restricted to this context. Multiobjective programming principles have broad applicability and might be appropriate for a variety of other MDS problems. We have demonstrated these principles for the case of finding a set of dimension permutations that provide good fits for each matrix in a set of dissimilarity matrices. However, multiobjective programming techniques can also be used to find a solution for a single matrix that provides a good fit as measured by two or more loss functions (or Stress measures in the case of nonmetric MDS). Our hope is that this chapter will stimulate further development of applications and methods for multiobjective MDS.

Michael J. Brusco
Florid State University

Stephanie Stahl
Tallahassee, FL

J. Dennis Cradit

Florid State University   .

REFERENCES

Arabie, P. (1973). Concerning Monte Carlo evaluations of nonmetric multidimensional scaling algorithms. *Psychometrika, 38*, 607–608.

Arabie, P. (1991). Was Euclid an unnecessarily sophisticated psychologist? *Psychometrika, 56*, 567–587.

Arabie, P., & Hubert, L. J. (1992). Combinatorial data analysis. *Annual Review of Psychology, 43*, 169–203.

Ashby, F. G., Maddox, W. T., & Lee, W. W. (1994). On the dangers of averaging across subjects when using multidimensional scaling or the similarity-choice model. *Psychological Science, 5*, 144–151.

Attneave, F. (1950). Dimensions of similarity. *American Journal of Psychology, 3*, 515–556.

Borg, I., & Leutner, D. (1983). Dimensional models for the perception of rectangles. *Perception & Psychophysics, 34*, 257–267.

Brusco, M. J. (2001). A simulated annealing heuristic for unidimensional and multidimensional (city-block) scaling of symmetric proximity matrices. *Journal of Classification, 18*, 3–33.

Brusco, M. J. (2002). Identifying a reordering of the rows and columns of multiple proximity matrices using multiobjective programming. *Journal of Mathematical Psychology, 46*, 731–745.

Brusco, M. J., Cradit, J. D., & Stahl, S. (2002). A simulated annealing heuristic for a bicriterion problem in market segmentation. *Journal of Marketing Research, 39*, 99–109.

Brusco, M. J., Cradit, J. D., & Tashchian, A. (2003). Multicriterion clusterwise regression for joint segmentation settings: An application to customer value. *Journal of Marketing Research, 40*, 225–234.

Brusco, M. J., & Stahl, S. (2001). An interactive approach to multiobjective combinatorial data analysis. *Psychometrika, 66*, 5–24.

Carroll, J. D., & Arabie, P. (1980). Multidimensional scaling. *Annual Review of Psychology, 31*, 607–649.

Carroll, J. D., & Arabie, P. (1998). Multidimensional scaling. In M. H. Birnbaum (Ed.), *Management, judgment, and decision making* (pp. 179–250). San Diego, CA: Academic Press.

Cho, R. Y., Yang, V., & Hallett, P. E. (2000). Reliability and dimensionality of judgments of visually textured materials. *Perception & Psychophysics, 62*, 735–752.

Davison, M. L. (1983). *Multidimensional scaling.* New York: Wiley.

Delattre, M., & Hansen P. (1980). Bicriterion cluster analysis. *IEEE Transactions on Pattern Analysis and Machine Intelligence, 2*, 277–291.

DeSarbo, W. S., & Grisaffe, D. (1998). Combinatorial optimization approaches to constrained market segmentation: An application to industrial market segmentation. *Marketing Letters, 9*, 115–134.

Dunn, J. C. (1983). Spatial metrics of integral and separable dimensions. *Journal of Experimental Psychology: Human Perception & Performance, 9,* 242–257.

Ferligoj, A., & Batagelj, V. (1992). Direct multicriteria clustering algorithms. *Journal of Classification, 9,* 43–61.

Furnas, G. W. (1989). Metric family portraits. *Journal of Classification, 6,* 7–52.

Garner, W. R. (1974). *The processing of information and structure.* Potomac, MD: Lawrence Erlbaum Associates.

Grau, J. W., & Nelson, D. K. (1988). The distinction between integral and separable dimensions: Evidence for the integrality of pitch and loudness. *Journal of Experimental Psychology: General, 117,* 347–370.

Groenen, P. J. F., & Heiser, W. J. (1996). The tunneling method for global optimization in multidimensional scaling. *Psychometrika, 61,* 529–550.

Groenen, P. J. F., Heiser, W. J., & Meulman, J. J. (1998). City-block scaling: Smoothing strategies for avoiding local minima. In I. Balderjahn, R. Mathur, & M. Schader (Eds.), *Classification, data Analysis, and data highways* (pp. 46–53). Berlin: Springer.

Groenen, P. J. F., Heiser, W. J., & Meulman, J. J. (1999). Global optimization of least-squares multidimensional scaling by distance smoothing. *Journal of Classification, 16,* 225–254.

Heiser, W. J. (1989). The city-block model for three-way multidimensional scaling, In R. Coppi & S. Belasco (Eds.), *Multiway data analysis* (pp. 395–404). Amsterdam: North-Holland.

Hoben, T. (1968). Spatial models and multidimensional scaling of random shapes. *American Journal of Psychology, 81,* 551–558.

Hubert, L. J., & Arabie, P. (1988). Relying on necessary conditions for optimization: Unidimensional scaling and some extensions. In H. H. Bock (Ed.). *Classification and related methods of data analysis* (pp. 463–472). Amsterdam: North-Holland.

Hubert, L., & Arabie, P. (1994). The analysis of proximity matrices through sums of matrices having (anti-) Robinson forms. *British Journal of Mathematical and Statistical Psychology, 47,* 1–40.

Hubert, L., Arabie, P., & Hesson-McInnis, M. (1992). Multidimensional scaling in the city-block metric: A combinatorial approach. *Journal of Classification, 9,* 211–236.

Hubert, L., Arabie, P., & Meulman, J. (1997). Linear and circular unidimensional scaling for symmetric proximity matrices. *British Journal of Mathematical and Statistical Psychology, 50,* 253–284.

Hubert, L., Arabie, P., & Meulman, J. (2001). *Combinatorial data analysis: Optimization by dynamic programming.* Philadelphia: Society for Industrial and Applied Mathematics.

Hubert, L. J., & Baker, F. B. (1977). The comparison and fitting of given classification schemes. *Journal of Mathematical Psychology, 16,* 233–253.

Hubert, L. J., & Busk, P. (1976). Normative location theory: Placement in continuous space. *Journal of Mathematical Psychology, 14,* 187–210.

Krieger, A. M., & Green, P. E. (1996). Modifying cluster-based segments to enhance agreement with an exogenous response variable. *Journal of Marketing Research, 33,* 351–363.

Kruschke, J. K. (1993). Human category learning: Implications for backpropagation models. *Connection Science, 5,* 3–36.

Kruskal, J. B. (1964a). Multidimensional scaling by optimizing goodness of fit to a nonmetric hypothesis. *Psychometrika, 29,* 1–27.

Kruskal, J. B. (1964b). Nonmetric multidimensional scaling: A numerical method. *Psychometrika, 29,* 115–128.

Lawson, C. L., & Hanson, R. J. (1974). *Solving least squares problems.* Englewood Cliffs: Prentice-Hall.

Lee, M. D. (2001). Determining the dimensionality of multidimensional scaling representations for cognitive modeling. *Journal of Mathematical Psychology, 45,* 149–166.

Lee, M. D., & Pope, K. J. (2003). Avoiding the dangers of averaging across subjects when using multidimensional scaling. *Journal of Mathematical Psychology, 47,* 32–46.

MacKay, D. B. (2001). Probabilistic multidimensional scaling using a city-block metric. *Journal of Mathematical Psychology, 45,* 249–264.

Melara, R. D. (1989). Similarity relations among synesthetic stimuli and their attributes. *Journal of Experimental Psychology: Human Perception and Performance, 15,* 212–231.

Monohan, J. S., & Lockhead, G. R. (1977). Identification of integral stimuli. *Journal of Experimental Psychology: General, 106,* 94–110.

Morgan, B. J. T., Chambers, S. M., & Morton, J. (1973). Acoustic confusion of digits in memory and recognition. *Perception & Psychophysics, 14,* 375–383.

Myung, I. J., & Shepard, R. N. (1996). Maximum entropy inference and stimulus generalization. *Journal of Mathematical Psychology, 40,* 342–347.

Nosofsky, R. M. (1992). Similarity scaling and cognitive process models. *Annual Review of Psychology, 43,* 25–53.

Rao, A. R., & Lohse, G. L. (1996). Towards a texture naming system: Identifying relevant dimensions of texture. *Vision Research, 36,* 1649–1669.

Ronacher, B. (1992). Pattern recognition in honeybees: Multidimensional scaling reveals a city-block metric. *Vision Research, 32,* 1837–1843.

Shepard, R. N. (1962a). The analysis of proximities: Multidimensional scaling with an unknown distance function. I. *Psychometrika, 27,* 125–140.

Shepard, R. N. (1962b). The analysis of proximities: Multidimensional scaling with an unknown distance function. II. *Psychometrika, 27,* 219–246.

Shepard, R. N. (1964). Attention and the metric structure of stimulus space. *Journal of Mathematical Psychology, 1,* 54–87.

Shepard, R. N. (1974). Representation of structure in similarity data: Problems and prospects. *Psychometrika, 39,* 373-422.

Shepard, R. N. (1987). Toward a universal law of generalization for psychological science. *Science, 237,* 1317–1323.

Shepard, R. N., & Arabie, P. (1979). Additive clustering: Representation of similarities as combinations of discrete overlapping properties. *Psychological Review, 86,* 87–123.

Shepard, R. N., & Cermak, G. W. (1973). Perceptual-cognitive explorations of a toroidal set of free-form stimuli. *Cognitive Psychology, 4,* 351–377.

Siegler, R. S. (1987). The perils of averaging data over strategies: An example from children's addition. *Journal of Experimental Psychology: General, 116,* 250–264.

Suppes, P., Krantz, D. M., Luce, R. D., & Tversky, A. (1989). *Foundations of measurement*, Vol. II, *Geometrical, threshold, and probabilistic representations.* New York: Academic Press.

Tversky, A., & Gati, I. (1982). Similarity, separability, and the triangle inequality. *Psychological Review, 89,* 123–154.

Vega-Bermudez, F., Johnson, K. O., & Hsiao, S. S. (1991). Human tactile pattern recognition: Active versus passive touch, velocity effects, and patterns of confusion. *Journal of Neurophysiology, 65,* 531–546.

Wuerger, S. M., Maloney, L. T., & Krauskopf, J. (1995). Proximity judgments in color space: Tests of a Euclidean color geometry. *Vision Research, 35,* 827–835.

CHAPTER 5

CRITICAL DIFFERENCES IN BAYESIAN AND NON-BAYESIAN INFERENCE AND WHY THE FORMER IS BETTER

5.1 INTRODUCTION

In this chapter we provide an introduction to Bayesian statistics directed at research in the social sciences, with an emphasis on how this approach differs from traditional statistical procedures. We present the mechanics of Bayesian inference along with the underlying theoretical justification. Included is an illustrative application using real data.

There are three primary differences between Bayesian and non-Bayesians that are highlighted here. First, all unknown values are treated probabilistically in Bayesian statistics by assigning and updating an associated probability distribution. Second, since we always have unknown values (i.e., model parameters) at the start of the process, Bayesians need to stipulate such distributions prior to analyzing the data. This is both an encumbrance and a feature, as we will see. Third, inference proceeds in the Bayesian fashion by updating (i.e., conditioning) these prior distributions on the data at hand. Conversely, non-Bayesian inferential procedures assume total ignorance about the distribution of unknown quantities, other than their role in an assumed likelihood function, and produce the single value that is maximally likely to occur given a single data set. Frequentists and likelihoodists differ here primarily in the context of the data generation procedure. It is either from an infinitely reoccurring stream according to the same mechanism (frequentists), or it is a one-off entity (likelihoodists).

So why might one be especially interested in using the Bayesian approach just described in the social sciences? Consider the manner in which data typically come to us. Many sociologists, political scientists, economists, and others obtain high-reliability survey or other cross-sectional data sets from government sources, academic projects, or collaborative ventures. These *observational* data sets are almost always created in a particular time/place setting such that exact replication is impossible. Unfortunately, a key theoretical basis of the twentieth century statistical inference assumes that the data are plucked from an ongoing stream of independent, identically distributed (i.i.d.) values (i.e., frequentism). This is not only impossible in many settings, it is misleading in that surveys or other collection instruments that are repeated even days or weeks apart cannot be assumed to replicate human thinking or attitudes with replicable precision. It makes much more sense, then, to assume that the data are unique and fixed at this point in time.

The key difference, therefore, for the social sciences is that the underlying phenomena of interest are typically not fixed physical constants, but, rather, aggregated human characteristics. These are, by nature, fluid and probabilistic. This is not to say that for some particular circumstance in space and time the unknown parameter value is not a fixed value, but that generalizations should be treated more broadly. Even in the circumstances where we are focused enough to be interested in a fixed, unknown value, there are intuitive advantages to evaluating these quantities probabilistically to express our obvious uncertainty.

Another critical difference lies in the treatment of knowledge that exists before the data collection and modeling enterprise. Very few, if any, scientists approach the statistical component of the research process with no existing knowledge of the effects under study. Although this is sometimes very general or vague information, often it is relatively informed about phenomena of interest. So how do researchers include such information? One way is the required stipulation of a probabilistic statement, probability distribution function or probability mass function, to describe the data generation process. For instance, specifying a Poisson PMF distribution for data that counts events means making the following four assumptions.

1. The time interval of interest can be divided into arbitrarily many small subintervals.

2. The probability of an event remains constant throughout the time interval of study.

3. The probability of more than one occurrence in a small subinterval can be ignored.

4. Events are independent.

(These assumptions will actually apply to our running example in this chapter.) Now how many authors do we see justifying each of these assumptions! To be fair, most readers have read and accept them implicitly for common situations, although it is possible to name instances in the social sciences that violate each of these where the Poisson distribution is used anyway. Probably the most routinely violated of the four is the last (e.g., instances of coups tends to breed other instances of coups either within-country or across neighbors). However, specifying the PDF or PMF alone often does not fully provide the prior information at the researcher's hand. In the Poisson example, we might know with some confidence that the intensity parameter is small or large based on previous observations. If we are counting government dissolutions in Italy, a large value is expected, but if we are counting nuclear weapons tests by developing nations, a small value is expected. Such information, even with reliable justification, is ignored completely in standard parameter estimation in such cases.

So how do Bayesian differ here? As noted above, Bayesians *must* provide distributional information on unknown parameters at the start of the modeling process. This is exactly how the researcher's information is inserted into the specification. If one believes that a count of cabinet dissolutions in Italy is going to be relatively high based on the number of events since World War II, it makes sense to assert a prior distribution for the Poisson intensity parameter that has a relatively high mean. Furthermore, since this is a distribution, not a single point, it is possible to provide different views of uncertainty around this mean by adjusting the variance of the prior distribution given. This, it turns out, is a marvelous way to express limited knowledge about various research quantities. We can state what we know or believe and at the same time express relative uncertainty in rigorous and overt fashion.

Finally, the inference process differs in Bayesian statistics in substantial ways. The key underlying philosophy is that of *updating degrees of belief.* We start with a prior distribution, which can be as vague as we like, and then condition on the newly obtained data at hand. The result is a new probability statement that we will call a *posterior distribution* since it is produced after these data are accounted for in the model. But wait, since this is also a regular distributional statement in the standard Kolmogorov sense (Billingsley, 1986), it too can be treated as a prior distribution if an additional set of data arrives. So prior and posterior are just relative terms that apply to one iteration of updating based on a single data set and the process can continue as long as we like (or as long as relevant data arrive). This process conforms exactly to the definition of scientific progress: current knowledge produces, theories which are tested, and knowledge is updated for future purposes. Great, right? How about an opposing paradigm that assumes that anything worth studying cannot be dynamic (except in the time-series context), data are produced in the same fashion forever, and it is better to assume prior ignorance at the start of the modeling process?

5.2 THE MECHANICS OF BAYESIAN INFERENCE

So far we have been relatively loose about the actual process of building Bayesian models and making inferences, which will now be rectified. To begin with, two explicit probability statements (PMF or PDF) are required:

- • A probability model for the data (given the parameters of such model) which is used to construct a likelihood function for the given sample

- • Unconditional probability distributions for unknown parameters

So the data get a standard assumption, but we now provide specific parameter forms for the unknowns. There are several caveats here. First, missing data are treated like parameters and estimated, therefore requiring priors themselves (see Little and Rubin, 1983, 1987; Rubin, 1987; Schaefer, 1997). Second, while the prior distributions cannot be conditioned on the data hand (except with empirical Bayes, see Morris, 1983; and Carlin and Louis, 2001), they can be conditioned on each other. For instance in cases where the data are assumed to be normally distributed, an unconditional prior is commonly stipulated for σ^2, but for the mean, a conditional prior of the form $\mu|\sigma^2$ is necessary for useful computational properties. [4] Third, although this is a completely parametric process, nonparametric Bayes is an increasingly appealing alternative (Ferguson, 1973; Antoniak, 1974; Dalal and Hall, 1980; Liu, 1996).

So now that we have a likelihood function for the observed data and a set of priors for the unknown quantities, these priors are conditioned on the data by multiplication in an application of Bayes' law to obtain the posterior distribution, which is the joint distribution of the parameters of interest conditioned on the data. This proceeds as follows:

$$\pi(\theta|x) = \frac{p(\theta)L(\theta|x)}{\int_\Theta p(\theta)L(\theta|x)d\theta}$$
$$\propto p(\theta)L(\theta|x)$$

posterior probability $\propto$ prior probability $\times$ likelihood function

Note that proportionality is used here since the denominator has no inferential value for the parameter θ, but simply ensures that the posterior is standardized in the sense of summing or integrating to 1. This last issue can easily be solved by rescaling after the Bayes' law process, and it is often easier to disregard this part for now. So we see that Bayesian inference proceeds through the use of Bayes' law to obtain the inverse probability of interest, which is the joint posterior. The last step is to evaluate the fit of this model and to test its sensitivity to the choice of both priors and likelihood. The latter process usually involves comparing the posterior produced to alternatives that come from more diffuse or general prior forms. The model-checking process seems like an obvious step, but it is actually one where Bayesian standards are much more cautious than non-Bayesian processes, where this step is rarely considered.

---

[4]Specifically, this setup gives the *conjugacy* property where the prior and posterior distributions belong to the same distributional family form. This is very convenient and was once important to Bayesians since it readily produced an analytically tractable posterior form. The advent of modern Bayesian stochastic simulation (Markov chain Monte Carlo) has made conjugacy less important but still appealing.

5.2.1 Example with Count Data

As an example of this model-building process, consider a data set consisting of counts, for which we specify a Poisson likelihood function:

$$f(\mathbf{y}|\mu) = \left(\prod_{i=1}^{n} y_i!\right)^{-1} \exp\left[\log(\mu)\sum_{i=1}^{n} y_i\right]\exp[-n\mu] \qquad (5.1)$$

A convenient form of the prior for μ (the intensity parameter) is a gamma distribution $\mathcal{G}(\alpha, \beta)$ with prior parameters α and β which the researcher specifies:

$$f(\mu|\alpha, \beta) = \frac{1}{\Gamma(\alpha)}\beta^{\alpha}\mu^{\alpha-1}e^{-\beta\mu}, \qquad \mu, \alpha, \beta > 0 \qquad (5.2)$$

This is handy because the gamma distribution is *conjugate* to the Poisson likelihood in Bayesian analysis, meaning that the posterior is also a gamma form:

$$\pi(\mu|\mathbf{y}) \propto p(\mu|\alpha, \beta)L(\mathbf{y}|\mu)$$

$$= \frac{\beta^{\alpha}\mu^{\alpha-1}}{\Gamma(\alpha)}\exp(-\beta\mu)\left(\prod_{i=1}^{n} y_i!\right)^{-1}\exp\left[\log(\mu)\sum_{i=1}^{n} y_i\right]\exp[-n\mu]$$

$$\propto \mu^{\alpha-1}\exp(-\beta\mu)\exp\left[\log(\mu)\sum_{i=1}^{n} y_i\right]\exp[-n\mu]$$

$$\propto \mu^{(\alpha+n\bar{y})-1}\exp\left[-(\beta+n)\mu\right]. \qquad (5.3)$$

Therefore, the posterior distribution for μ is $\mathcal{G}(\alpha + n\bar{y}, \beta + n)$, and thus has mean $(\alpha + n\bar{y})/(\beta + n)$ and variance $(\alpha + n\bar{y})/(\beta + n)^2$. The neat part about this is that we now have everything we need to know about the posterior and can either rely upon known properties of the gamma distribution or simply simulate values according to this parameterization and summarize empirically.

As an illustration, consider counts of U.S. testing of thermonuclear devices up until the 1992 Test Ban Treaty, which the United States signed.[5] The first genuinely thermonuclear device, "Mike," was detonated November 1, 1952 on the former Islet of Elugelab (it apparently no longer exists) [see DeGroot (2004, pp. 175–8) for interesting details on the history and policy of testing in the United States], so the data set for hydrogen bombs is set to start in calendar year 1953. The data are given in Table 5.1

The data have mean $\bar{y} = 25.55$ over this period. Suppose that we set $\alpha = 50$ and $\beta = 2$, to be somewhat sympathetic with the data. Let us also set $\alpha = 5$ and $\beta = 1$ as a cynical prior to serve as a comparison specification and thus a test of how influential the first prior turns out to be. These priors and the associated posterior summaries are given in Table 5.2

---

[5] The data are available at http://artsci.wustl.edu/~jgill/replication.html.

Table 5.1 Counts of U.S. Thermonuclear Tests By Year

Year	Count	Year	Count	Year	Count	Year	Count	Year	Count
1953	11	1954	6	1955	18	1956	18	1957	32
1958	77	1959	0	1960	0	1961	10	1962	98
1963	47	1964	47	1965	39	1966	48	1967	42
1968	56	1969	46	1970	39	1971	24	1972	27
1973	24	1974	23	1975	22	1976	21	1977	20
1978	21	1979	16	1980	17	1981	17	1982	19
1983	19	1984	20	1985	18	1986	15	1987	15
1988	15	1989	12	1990	9	1991	8	1992	6

Source: U.S. Department of Energy.

Table 5.2 Gamma–Poisson Model for Thermonuclear Tests

	Prior		Posterior	
Prior Parameters	Mean	Variance	Mean	Variance
$\alpha = 50, \beta = 2$	25	12.5	25.52	0.60
$\alpha = 5, \beta = 1$	5	5.0	25.05	0.61

What this shows is that the posterior distribution is relatively insensitive to the form of the prior that we select. This robustness property is reassuring in that it demonstrates that we are letting the data speak clearly here. To further illustrate these results, consider Figure 5.1. We see here how dramatically "wrong" the second prior specification is, yet how it changes relatively little in the final analysis. The darker curve is the posterior distribution, and the lighter curve is the associated prior distribution.

So how would a non-Bayesian analysis of these data proceed? We know that the maximum likelihood estimate of the intensity parameter from a Poisson specification is the data mean, $\bar{y} = 25.55$, which differs little from the analysis above except that it buries the assumption of a uniform prior over the support of μ. More specifically, every non-Bayesian result is equivalent to a Bayesian result where the prior is an appropriately bounded uniform distribution. So non-Bayesians assume a priori that all possible outcomes are equally likely. This does not seem like a terrible assumption except that we *know* something about anticipated thermonuclear tests here. For instance, five is much more likely than 500, and so on. Many people can still live

Figure 5.1   Prior and posterior distributions for thermonuclear tests

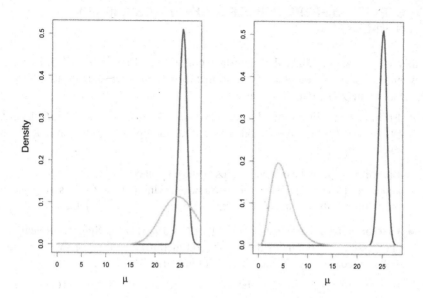

with such uniform assumptions, but the usual sin here is not stating this overtly as a model component, and therefore having to defend it.

5.2.2   Comments on Prior Distributions

So where do prior distributions come from? We see above that they are part of the Bayesian process, and the only practical guidance provided so far is, for the purpose of argument, to compare the chosen prior to a more cynical form. Prior distributions come from varied sources, such as: previous studies and published work, researcher intuition, information provided by substantive experts, convenience (i.e. conjugacy, vagueness), nonparametrics and other data sources, and those that give diagnostic information. Nothing limits the sources of these priors other than an ability for the author to justify what is used in a particular model. A standard rule here is that if pertinent information exists that affects the quantities of interest, it should be expressed in the prior. One would not think that the last sentence is controversial, but it remains so. The heart of this controversy is the idea of what is "objective" and what is "subjective." Although it may disappoint some, all statistical models are subjective in the sense that the data used are chosen rather than required, a likelihood function is also user-selected, and the form of testing is also a personal decision. Thus the prior simply joins this list as another subjective choice among many. Bayesians therefore prefer all these decisions to be overt, so "hiding" behind a veneer of objectivity does not make sense.

5.3 SPECIFIC DIFFERENCES BETWEEN BAYESIANS AND NON-BAYESIANS

In this section we highlight the important philosophical and practical differences between Bayesian methods and the traditional forms. First, let us delineate the three important branches of statistical reasoning.

- **Frequentists:** From the Neyman-Pearson (1928a,b, 1933a,b, 1936a,b) and Wald (1950) setup. An orthodox view that sampling is infinite and decision rules can be sharp.

- **Bayesians:** From the original Bayes (1763), Laplace (1774, 1781, 1811, 1814) and de Finetti (1972, 1974, 1975) tradition. Unknown quantities are treated probabilistically and the state of the world can always be updated.

- **Likelihoodists:** From Fisher's (1925a, 1934, 1955) basis. Single sample inference based on maximizing the likelihood function and relying on the Birnbaum (1962) theorem. Bayesians who don't know that they are.

Thus, the important distinction is between the frequentist approach and the Bayesian paradigm, since Fisher's idea of using the likelihood as a data reduction device merely ignores the associated uniform prior. So for the time-being we concentrate on the greater contrast between Bayesian inference and Neyman–Pearson testing before we move on to the null hypothesis significance test. The summary of the main differences discussed in this section is given in Table 5.3 For further discussions of such differences, see the excellent recent essay by Little (2006) and the classic discussion by Efron (1986).

The first important difference relates to what is considered *fixed* in the model. In frequentist inference the data are random in the form of an iid sample from a continuous stream and the parameters are fixed (but generally unknown) by nature. Conversely, the Bayesian treats data as fixed since they are observed and therefore non-changing in the analysis, and the unknown parameters are considered random to reflect uncertainty. Since the unknown parameters are considered random unknown quantities, they are described with distributional forms conditional on the most recently acquired information.

Surprisingly, Bayesians and frequentists actually differ on the definition of probability. To a frequentist, probability is just the long-run proportion of times that some event occurs in a controlled setting that assures iid sampling, which comes from their fundamental idea of replicability of the data generation process. This means that the probabilistic quantity of interest is the distribution of these data under various hypotheses. On the other hand, Bayesians see probability as the researcher/observer "degree of belief" before or after the data are observed. So a Bayesian view of probability is personal and represents an individual willingness to gamble given some payout.

Table 5.3 Critical Differences Between Inference Paradigms

	Bayesian treatment	Frequentist treatment
Starting point	$p(\theta)$	Feigned ignorance
Fixed quantity	Data	Unknown parameters
Stochastic quantity	Unknown parameters	Data
Focus of interest	$p(\theta\|data)$	$p(data\|H_0)$
Inference tool	Probabilistic description of $p(\theta\|data)$	Deduction from $p(data\|H_0)$, by setting α in advance
Summarization	HPD intervals and probability statements	Acceptance rules based on $p(data\|H_0) < \alpha$

Inference differs dramatically, as we will see in detail below. Frequentists use point estimates for unknown parameters along with standard errors or 95% *confidence* intervals. Deduction is then from $p(data|H_0)$, by setting α in advance. The rule is to accept H_1 if $p(data|H_0) < \alpha$, and accept H_0 if $p(data|H_0) \geq \alpha$. These are firm rules. A Bayesian eschews hard-and-fast rules in favor of finding the weight of evidence in some direction. Induction comes from $p(\theta|data)$, starting with $p(\theta)$, and broad descriptions of the posterior distribution such as means and quantiles constitute scientific evidence in some direction. Often, highest posterior density intervals, indicating region of highest posterior probability regardless of contiguity, are used to summarize most likely values of the unknown parameters.

Both frequentists and Bayesians perform post-hoc model quality checks to increase reliability. Frequentists typically use calculation of type I and type II errors, setting α in advance. Sometimes effect size and/or power are calculated, although much more rarely in the social sciences. Usually, there is just fixation with small differences in p-values despite large measurement error in the social sciences relative to other scientific disciplines. Bayesian tools here include: posterior predictive checks from integrating over posterior, sensitivity checks to forms of the prior, and more formal tools for model comparison such as the Bayes factor, Bayesian information criterion (BIC), and the more recent deviance information criterion (DIC).

5.4 PARADIGMS FOR TESTING

Our purpose here is actually not to contrast canonical frequentism with Bayesian approaches since few, if any, social science researchers practice the pure form of Neyman–Pearson (1928a,b, 1933a,b, 1936a,b) testing. Instead, a strange animal called the *null hypothesis significance test* (NHST) dominates. The social sciences are burdened with this synthesis of the Fisher's *test of significance* and Neyman and Pearson's *hypothesis test*. In the NHST hybrid procedure, two hypotheses are posited: a null or restricted hypothesis, H_0, which competes with an alternative or research hypothesis, H_1, describing two complementary notions about some unknown parameter of interest. The researcher's hypothesis is the model that asserts some nonzero effect and is operationalized through statements about the associated parameter, β here. Most commonly, the null asserts that $\beta = 0$, and the complementary research hypothesis is that $\beta \neq 0$. This fits well with standard regression models, even though more generally, the test evaluates a parameter vector: $\beta = \{\beta_1, \beta_2, \ldots, \beta_m\}$, and the null hypothesis places ℓ restrictions on some subset ($\ell \leq m$) of the vector, such as $\beta_i = k_1\beta_j + k_2$ with constants k_1 and k_2.

The NHST uses a test statistic (B), which is some function of β and the data, and is calculated and compared with its known distribution under the assumption that H_0 is true. Typical test statistics are sample means ($\bar{X}$), χ^2 statistics from contingency tables, and t-statistics from linear or nonlinear regression. The test procedure then assigns one of two decisions (D_0, D_1) to all possible values in the sample space of B, each supporting either H_0 or H_1 respectively. The p-value is equal to the area in the tail (or tails) of the assumed distribution under H_0 which starts at the point designated by the placement of B on the horizontal axis and continues to infinity. A predetermined α level is usually not specified and the p-value is, instead, positioned on a scale between Fisher's convenient cut-offs, according to the categories: $[\{> 0.1\}, \{0.1 : 0.05\}, \{0.05 : 0.01\}, \{< 0.01\}]$, with great joy produced from placement toward the right of this spectrum. Stars (asterisks) are sometimes given in place of these intervals. This test, as described, is really an accidental hybrid between Fisher's test of significance and Neyman and Pearson's hypothesis test, without any known author or creator.

Fisher (1925a, 1934, 1955) gives a single hypothesis, H_0, with a known distribution of the test statistic B. The further this test statistic is away from its expected value under the null, the less plausible H_0 becomes. The location of the test statistic establishes density in the tail or tails of the distribution, the size of which is the p-value. So the Fisher steps are:

- Specify the distribution implied by the null hypothesis.

- Fix the test statistic and its distribution under the assumption that the null hypothesis is true.

- Calculate this test statistic from the data.

- Determine the p-value level that corresponds to the test statistic.

- Reject H_0 if this p-value is sufficiently small.

- Otherwise, reach no conclusion.

The obvious question that remains is what constitutes a sufficiently small p-value as to reject the null. Fisher gives us the familiar thresholds but was surprisingly flexible about choosing levels: "The value for Q is therefore significant on the higher standard (1 per cent) and that for N_2 at the lower standard (5 per cent)." (1971, pp.152–153).

Neyman and Pearson (1928a,b, 1933b, 1936a) do not follow Fisher's supposition that only the null hypothesis needs to be tested, and they propose two complementary hypotheses: Θ_A and Θ_B, neither of which needs to be a "null" in Fisher's sense. Furthermore, in this construct it is possible to specify multiple Θ_B alternatives against a single Θ_A. With two competing hypotheses in any one test, Neyman and Pearson define an a priori selected α, the probability of falsely rejecting Θ_A under the assumption that it is true (type I error), and β, the probability of failing to reject Θ_A when it is false (type II error). It is important to remember that α and β are probabilities *conditional* on two mutually exclusive events: α is conditional on Θ_A being true, and β is conditional on Θ_A being false. Also, Neyman and Pearson (1933a, 1936a) call $1 - \beta$ the *power* of the test: the long-run probability of accurately rejecting a false null hypothesis given a point alternative hypothesis, and they seek tests with the highest possible power for a given sample and desired significance level chosen in advance. So the Neyman–Pearson steps are:

- Identify the hypothesis of interest, Θ_B, and its complement, Θ_A.

- Determine the test statistic and its distribution under the assumption that Θ_A is true.

- Specify a significance level.

- Determine the corresponding critical value of the test statistic under the assumption that Θ_A is true.

- Calculate the test statistic from the data.

- Reject Θ_A and accept Θ_B if the test statistic is further than the critical value from the expected value of the test statistic.

- Otherwise accept Θ_A.

Note that the level α is chosen in advance and applied rigidly. For instance, if .05 is desired as a long-run probability, it is necessary to reject Θ_A for an achieved significance level of .0499 and accept Θ_A for an achieved significance of .05001. This dogmatism surprises many people and does not seem to fit well in the social sciences, where measurement error typically exceeds any such distinctions.

The null hypothesis significance test as practiced in the social sciences tries to blend these two approaches, but produces an inconsistent product with several pathologies. It uses Fisher's null idea but also includes a specific alternative, which he did not advocate. Fisher has no definition of the power of the test nor of accepting alternative hypotheses in the final interpretation, whereas Neyman and Pearson state that rejection of one implies acceptance of the other, and this rejection is based on a predetermined α level. The NHST has no provision for acceptance, and graduate students are indoctrinated to say things like "there is evidence in support of the hypothesis...." Furthermore, Neyman and Pearson's hypothesis test defines the significance level a priori as a function of the test, but Fisher's test of significance defines the significance level afterwords as a function of the data. The NHST straddles this difference by pretending to select α in advance, but actually binning p-values into categories as strength of evidence. Thus there is an implied alternative hypothesis but no determination of power. The result is confusion about long-run probabilities and the definition of the p-value. Since most social scientists do not perform repeated controlled experiments, the probability of a type I error does not constitute a long-range probability of rejection. The pathologies that emerge are truly damaging: a logical inconsistency coming from probabilistic modus tollens, confusion over the order of the conditional probability, chasing significance but ignoring effect size, adherence to the completely arbitrary significance thresholds, and confusion about the probability of rejection. An excellent recent essay on these issues can be found in Denis (2005). More important, all of these issues are avoided with Bayesian inference, which suffer from none of the foregoing problems or misinterpretations.

The bottom line is that the pseudo-frequentist NHST is not just suboptimal; it is actually wrong. More than a few authors have noted this. For reference, consider a small sample of these works: Barnett, 1973; Berger, Boukai, & Wang, 1997; Berger & Sellke, 1987; Berkhardt & Schoenfeld, 2003; Bernardo, 1984; Carver, 1978, 1993; Cohen, 1962, 1977, 1988, 1992, 1994; Denis, 2005; Falk & Greenbaum, 1995; Gelman, Carlin, Stern, & Rubin, 2003; Gigerenzer, 1987, 1993, 1998; Gigerenzer & Murray, 1987; Gill, 1999, 2007; Gliner, Vaske & Morgan, 2004; Grayson, 1998; Greenwald, 1975; Greenwald, Gonzalez, Harris & Guthrie, 1996; Hager, 2000; Howson & Urbach, 1993; Hunter, 1997; Hunter & Schmidt, 1990; Jeffreys, 1961; Kirk, 1996; Krueger, 2001; Lecoutre, Lecoutre & Poitevineau 2001; Lindsay, 1995; Loftus, 1991, 1993a,b, 1994, 1996; Loftus & Bamber, 1990; López, 2003; Macdonald, 1997; Meehl, 1967, 1978, 1990; Menon, 1993; Nickerson, 2000; Oakes, 1986; Pollard, 1993; Pollard & Richardson, 1987; Robinson & Levin, 1997; Rosnow & Rosenthal, 1989; Rozeboom, 1960; Schmidt, 1996; Schmidt & Hunter, 1977; Sedlmeier & Gigerenzer, 1989; Thompson, 2002; Wilkinson, 1977; Yoccoz, 1991.

Take, for instance, the idea of confidence intervals derived from the NHST. The basis for confidence intervals comes purely from the Neyman-Pearson construct since replicability is implied. As an illustration, which of these is the correct interpretation of a $(1 - \alpha)$ confidence interval?

- An interval that has a $1 - \alpha\%$ chance of containing the true value of the parameter.

- An interval that over $1 - \alpha\%$ of replications contains the true value of the parameter, *on average*.

The correct response is the second one. That is, a 95% CI covers the true parameter 19 times out of 20. However, even though the first response is incorrect, this is the interpretation that people *really want*. In fact, the first response is the interpretation of the Bayesian *credible interval* since the parameter is described probabilistically. This is the big distinction between Bayesian and non-Bayesian approaches: All Bayesian hypothesis testing produces probability statements about the parameters of interest or the hypotheses. Therefore, there are no pathologies related to repeated trials, the interpretation of tail values, or a priori versus posthoc determination of significance levels.

For instance, suppose that we want to test various hypotheses about the first posterior distribution from the U.S. thermonuclear testing example. This posterior was found to be $\mathcal{G}(\alpha + n\bar{y}, \beta + n)$. With $\alpha = 50, \beta = 2, n = 40$, and $\bar{y} = 25.25$, this is $\mathcal{G}(1072, 42)$. Knowing this, we can calculate analytically any desired summary statistic. It turns out be much easier to produce a summary by simulation (simulation is actually what set the Bayesians free, as we will see in the next section). A single line of R code produces a useful summary:

```
quantile(rgamma(100000,shape=1072,rate=42),c(.01,.1,.25,.5,.75,.9,.99))
     1%       10%       25%       50%       75%       90%       99%
23.74844 24.52746 24.99852 25.51865 26.04424 26.52773 27.37058
```

Now suppose that we want to test the hypothesis that the unknown intensity parameter is less than 24 against a complement:

$$H_0 : \mu \geq 24 \quad \text{versus} \quad H_1 : \mu < 24$$

Thus we want the posterior probability that μ is less than 24, $\pi(\mu < 24|\mathbf{y})$. Again this can be performed analytically but it is easier to use R:

```
pgamma(24,shape=1072,rate=42)
[1] 0.02358382
```

So we have a very low probability that H_1 is true and therefore a very large probability that H_0 is true. Notice how much cleaner this is than the NHST. Every statement is based on probabilities, no hypothesis holds a special asymmetrical position like the null in the NHST, and (credible) intervals based on the quantiles above are interpreted the intuitive way that people prefer.

Models themselves are also described probabilistically, so we can compute the ratio of model probabilities as a test of alternative specifications, M_1 versus M_2 using the same data $\mathbf{y}$. This is called the Bayes factor, and is calculated according to

$$B(\mathbf{y}) = \frac{\pi(M_1|\mathbf{y})/p(M_1)}{\pi(M_2|\mathbf{y})/p(M_2)} = \frac{\int_{\theta_1} f_1(\mathbf{y}|\theta_1)p_1(\theta_1)d\theta_1}{\int_{\theta_2} f_2(\mathbf{y}|\theta_2)p_2(\theta_2)d\theta_2} \quad (5.4)$$

where $P(M_i)$ is the prior probability put on the ith *model* with parameter vector θ_i (Gill, 2007). Very large $B(\mathbf{y})$ numbers imply strong support for the first model, and very small $B(\mathbf{y})$ numbers imply strong support for the second model, but values near 1 give inconclusive answers [see the categorization in Jeffreys (1961, p. 432) and Kass and Raftery (1995, p. 777)]. The Bayes factor is actually a generalization of the likelihood ratio that incorporates prior information and does not require nesting. Suppose that we run a Bayes factor test of the two models summarized in Table 5.2 Let the model with priors [$\alpha = 50, \beta = 2$] be M_1 and the model with priors [$\alpha = 5, \beta = 1$] be M_2. From (5.3), we get a Bayes factor for model 1 over model

2 (with equal model prior probabilities) from

$$
\begin{aligned}
B(\mathbf{y}) &= \frac{\int_M f_1(\mathbf{y}|\mu)p_1(\mu_1)d\mu}{\int_M f_2(\mathbf{y}|\mu)p_2(\mu_2)d\mu} \\
&= \frac{\int_0^\infty \frac{1}{\Gamma(\alpha_1)}\beta_1^{\alpha_1}\mu^{\alpha_1-1}\exp(-\beta_1\mu)\left(\prod_{i=1}^n y_i!\right)^{-1}}{\int_0^\infty \frac{1}{\Gamma(\alpha_2)}\beta_2^{\alpha_2}\mu^{\alpha_2-1}\exp(-\beta_2\mu)\left(\prod_{i=1}^n y_i!\right)^{-1}} \cdots \\
&\cdots \frac{\exp\left[\log(\mu)\sum_{i=1}^n y_i\right]\exp[-n\mu]d\mu}{\exp\left[\log(\mu)\sum_{i=1}^n y_i\right]\exp[-n\mu]d\mu} \\
&= \frac{\Gamma(\alpha_1+n\bar{y})\,\beta_1^{\alpha_1}\,(\beta_2+n)^{\alpha_2}}{\Gamma(\alpha_2+n\bar{y})\,\beta_2^{\alpha_2}\,(\beta_1+n)^{\alpha_1}}\left(\frac{\beta_2+n}{\beta_1+n}\right)^{n\bar{y}} \\
&= 612,049.7 \qquad\qquad\qquad\qquad\qquad\qquad (5.5)
\end{aligned}
$$

which gives obvious support to model 1. Bayes factors can be more difficult to calculate than that in this example, but a host of simulation and approximation tools help and the related Bayesian information criterion is computationally simple (Kass and Raftery, 1995).

5.5 CHANGE-POINT ANALYSIS OF THERMONUCLEAR TESTING DATA

Returning to the example of U.S. thermonuclear tests, we can easily notice that the later years have sharply reduced testing relative to the first two-thirds (roughly) of the data. The question that arises is: When did a change occur? We could do the appropriate historical research looking for treaties or changes in technology, but even armed with this information we cannot necessarily make a firm determination of the change year. Is there a fixed year that constitutes a firm change point? Probably not. So it seems logical to put a distribution on this change point to reflect a likely era of change. Thus the Bayesian model is substantively appropriate here. The point of this example is to demonstrate how some problems are naturally amenable to Bayesian analysis where non-Bayesian methods can be awkward.

The objective is to use this test data sequence to estimate the change point and also to obtain posterior estimates of the two Poisson intensity parameters since we hereby claim two eras. Now $y_1, y_2, \ldots, y_n$ are a series of count data where we hypothesize the existence of a change point at some year, k, along the series. This defines the two Poisson data-generating processes:

$$
\begin{aligned}
x_i|\lambda &\sim \mathcal{P}(\lambda) \quad i = 1, \ldots, k \\
x_i|\phi &\sim \mathcal{P}(\phi) \quad i = k+1, \ldots, n
\end{aligned}
$$

where the determination of which to apply depends on the location of the change point k. This means that there are three parameters to estimate: λ, ϕ, and k, and the different role of k makes this estimation process more challenging. The non-Bayesian approach has an awkward joint likelihood function in terms of maximization.

Our modeling exercise stipulates three independent priors:

$$\lambda \sim \mathcal{G}(\alpha, \beta)$$
$$\phi \sim \mathcal{G}(\gamma, \delta)$$
$$k \sim \text{discrete uniform on} [1, 2, \ldots, n]$$

and the prior parameters are assigned according to $\alpha = 1, \beta = 1, \gamma = 1, \delta = 1$. These values deliberately disadvantage the estimation process by asserting no change and a small number of values, and varying them produces little change in the resulting posterior distribution. Thus, we have a skeptical starting point, and reliable findings implies a strong statement by the data and therefore little concern that the priors are selected to be influential. For details on these models, see Gill (2007). From this we get the following joint posterior and its proportional simplification:

$$\pi(\lambda, \phi, k|y) \propto L(\lambda, \phi, k|y) \pi(\lambda|\alpha, \beta) \pi(\phi|\gamma, \delta) \pi(k)$$

$$= \left(\prod_{i=1}^{k} \frac{e^{-\lambda} \lambda^{y_i}}{y_i!} \right) \left(\prod_{i=k+1}^{n} \frac{e^{-\phi} \phi^{y_i}}{y_i!} \right) \left(\frac{\beta^\alpha}{\Gamma(\alpha)} \lambda^{\alpha-1} e^{-\beta\lambda} \right) \cdots$$

$$\times \cdots \left(\frac{\delta^\gamma}{\Gamma(\gamma)} \phi^{\gamma-1} e^{-\delta\phi} \right) \frac{1}{n}$$

$$\propto \lambda^{\alpha-1+\sum_{i=1}^{k} y_i} \phi^{\gamma-1+\sum_{i=k+1}^{n} y_i} e^{-(k+\beta)\lambda - (n-k+\delta)\phi}.$$

This turns out to be a difficult estimation problem (Bayesian or non-Bayesian), but if we can express this joint distribution as a set of full conditional distributions, we can use modern Bayesian stochastic simulation. The Gibbs sampler (Gelfand & Smith 1990) iterates through these conditional distributions until the resulting Markov chain converges to the limiting distribution. Thus, we replace a difficult integration process with summarizing empirical draws from $\pi(\lambda, \phi, k|y)$. The associated full conditional distributions for λ and ϕ are

$$\lambda|\phi, k \sim \mathcal{G}(\alpha + \sum_{i=1}^{k} y_i, \beta + k) \qquad (5.6)$$

$$\phi|\lambda, k \sim \mathcal{G}(\gamma + \sum_{i=k+1}^{n} y_i, \delta + n - k) \qquad (5.7)$$

and with greater work (Gill, 2007),

$$p(k|y) = \frac{f(y, \phi) L(y|k) p(k)}{\sum_{\ell=1}^{n} f(y, \phi) L(y|k_\ell) p(k_\ell)}$$

$$= \frac{L(y|k) p(k)}{\sum_{\ell=1}^{n} L(y|k_\ell) p(k_\ell)}. \qquad (5.8)$$

where $L(y|k) = \exp[k(\phi - \lambda)] (\lambda/\phi)^{\sum_{i=1}^{k} y_i}$ [i.e. λ and ϕ were suppressed from the notation only to make (5.8) visually clear]. Therefore, each iteration of the Gibbs sampler will calculate an n-length probability vector for k and draw a value accordingly. The Gibbs sampler is run for 1 million iterations, discarding the first 500,000 as a conservative burn-in period. Clearly, these are overly cautious numbers, but the algorithm runs very quickly in R. All standard diagnostics support the conjecture of convergence of the Markov chain, and the trace plots for these last 500,000 values are shown in Figure 5.2. Empirical summaries of these last 500,000 values are provided in Table 5.4

Table 5.4 Testing Change point Analysis

Quantile	λ	ϕ	k
Minimum	26.62	14.40	16.00
First quartile	31.94	18.94	18.00
Median	32.87	19.83	19.00
Third quartile	33.82	20.73	20.00
Maximum	39.03	25.42	26.00
Mean	32.89	19.83	19.36

First notice that the posterior median (and the rounded posterior mean) indicate that the center of the posterior distribution for the change point (k) is the year 1971. Notice that this is probably the year that one would guess for a change since there were 46 and 39 tests in 1969 and 1970, respectively, and 24 and 27 tests in 1971 and 1972, where these patterns are typical in those directions. Also, the posterior mean for the intensity parameter of the first period is noticeably larger than that for the second period, as expected. So it appears that the early 1970s saw a change of testing regimes for the United States. It turns out that the United States and 47 other nations, including Great Britain and the Soviet Union, signed the *Treaty for the Nonproliferation of Nuclear Weapons* in 1970. Shortly thereafter, the *Comprehensive Test Ban Treaty* was proposed, which would have the world's nuclear powers cease testing completely. Although it was not implemented at the time, its introduction may have had a chilling effect on testing procedures in the United States.

Consider the difficulty in approaching this problem from a non-Bayesian perspective. Maximizing the joint likelihood function,

$$\left(\prod_{i=1}^{k} \frac{e^{-\lambda}\lambda^{y_i}}{y_i!} \right) \left(\prod_{i=k+1}^{n} \frac{e^{-\phi}\phi^{y_i}}{y_i!} \right)$$

is not straightforward, due to the position of k in the two products. Common solutions involve making suspect normality assumptions (Sen & Srivastava 1975a,b), ãûøòï

Figure 5.2   Traceplots for U.S. thermonuclear testing sampler.

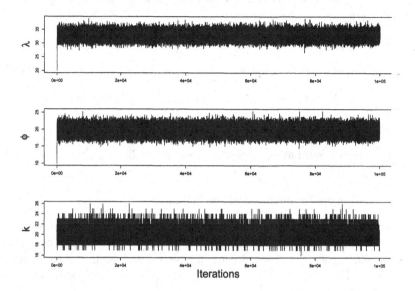

nonparametric models (Pettitt, 1979) and cumulative sums (Page, 1957; McGilchrist, and Woodyer 1975). The central problem in this context is testing the null hypothesis of no change point versus all possible change-point alternatives when the pertinent distribution is unknown. This explains why change-point problems are now gener- ally tackled from either a Bayesian perspective or using more complex time-series tools to allow serial dependence and the accompanying assumptions that they imply [Darkhovsky, 1994; Brodsky and Darkhovosky, 1993; Woodward and Gray, 1993; see also the nice annotated bibliography of Shaban (1980)]. Complications include allowing multiple change points, admitting an unknown number of change points, multidimensional outcome variables, nonconstant hazards, and the inclusion of co- variates.

5.6 CONCLUSION

In this brief chapter, we have outlined the basics of Bayesian statistical methodology and noted some important distinctions from classical statistical approaches. Our purpose is to persuade quantitative social scientists that the Bayesian view is not only superior; it is inevitable. The obvious cost is additional formalism and more complicated calculations, due to the inclusion of prior distributions and probabilistic interpretations. However, a host of computational tools have emerged to handle such problems, and since 1990, Bayesian models have not had any serious restrictions in producing and summarizing models. Bayesian stochastic simulation is used here in the running example to highlight the freedom that MCMC now provides.

As a last sales pitch, consider the following description of Bayesian procedures. Resulting models are characterized by:

- Direct statement model assumptions

- Rigorous *probability* statements about the quantities of theoretical interest

- The ability to update inferential statements (i.e. learn) as new data are received

- Systematic incorporation of qualitative or narrative knowledge

- Straightforward assessment of both model quality and sensitivity to assumptions

- Summary statements based on probabilities

We have noted that typical social science statistical reporting is deficient on every one of these items. It is curious, therefore, that non-Bayesian thinking and the deeply flawed null hypothesis significance test have dominated empirical social science. To be fair, we should also list reasons to eschew Bayesian approaches in the social sciences. These include:

- Population parameters are truly fixed and unchanging under all realistic circumstances

- Researchers do not have any information prior to model development

- Results need to be reported as if data were from a controlled experiment

- Statistical "significance" is more important than effect size

- Computers and other tools are either slow or unavailable

- Automated, "cookbook"-type procedures are preferred to individualized analysis

Hopefully, these points will prove to be thought-provoking to introspective researchers who might not otherwise have considered the underlying paradigm that they rely upon in regular empirical work.

Finally, the Bayesian view condenses down to three philosophical principles that adherents believe. First, the world should be viewed probabilistically, either because the subject of inquiry actually has a distribution or because uncertainty has a natural description via distributions. Second, every statistical model ever created in the history of the human race is subjective, so it is better to admit this and defend all model assumptions. Third, prior information abounds in the social sciences and it is important and helpful to use it. None of these ideas seems particularly controversial in the course of social science research, and the purpose of this chapter has been to introduce the associated procedures along with contrasts to their non-Bayesian alternatives.

Jeff Gill
Washington University, St. louis

REFERENCES

Antoniak, C. E. (1974). Mixtures of Dirichlet processes with applications to Bayesian non-parametric problems. *Annals of Statistics, 2*, 1152–1174.

Barnett, V. (1973). *Comparative statistical inference.* New York: Wiley.

Bayes, T. (1763). An essay towards solving a problem in the doctrine of chances. *Philosophical Transactions of the Royal Society of London, 53*, 370–418.

Berger, J. O., Boukai, B., & Wang, Y. (1997). Unified frequentist and Bayesian testing of a precise hypothesis. *Statistical Science, 12*, 133–160.

Berger, J. O., & Sellke, T. (1987). Test of a point null hypothesis: The irreconcilability of significance levels and evidence. With discussion. *Journal of the American Statistical Association, 82*, 112–139.

Berkhardt, H., & Schoenfeld, A. H. (2003). Improving educational research: Toward a more useful, more influential, and better-funded enterprise. *Educational Researcher, 32*, 3–14.

Bernardo, J. M. (1984). Monitoring the 1982 Spanish socialist victory: A Bayesian analysis. *Journal of the American Statistical Association, 79*, 510–515.

Billingsley, P. (1986). *Probability and measure.* New York: Wiley.

Birnbaum, A. (1962). On the foundations of statistical inference. *Journal of the American Statistical Association, 57*, 269–306.

Brodsky, B. E., & Darkhovsky, B. W. (1993). *Nonparametric methods in change point problems.* Dordrecht, The Netherlands: Kluwer.

Carlin, B. P., & Louis, T. A. (2001). *Bayes and empirical Bayes methods for data analysis.* 2nd ed. New York: Chapman & Hall.

Carver, R. P. (1978). The case against statistical significance testing. *Harvard Education Review, 48*, 378–399.

Carver, R. P. (1993). The case against statistical significance testing, revisited. *Journal of Experimental Education, 61*, 287–292.

Cohen, J. (1962). The statistical power of abnormal-social psychological research: A review. *Journal of Abnormal and Social Psychology, 65*, 145–153.

Cohen, J. (1977). *Statistical power analysis for the behavioral sciences.* New York: Academic Press.

Cohen, J. (1988). *Statistical power analysis for the behavioral sciences.* 2nd ed. Hillsdale, NJ: Lawrence Erlbaum Associates.

Cohen, J. (1992). A power primer. *Psychological Bulletin, 112*, 115–159.

Cohen, J. (1994). The Earth is round ($p < .05$). *American Psychologist, 12*, 997–1003.

Dalal, S. R., & Hall, G. J., Jr. (1980). On approximating parametric Bayes models by non-parametric Bayes models. *Annals of Statistics, 8*, 664–772.

Darkhovsky, B. S. (1994). Nonparametric methods in change-point problems: A general approach and some concrete algorithms. In E. Carlstein, H. Möller, & D. Siegmund (Eds.), *Change-point problems*, pp.99–107. Hayward, CA: Institute of Mathematical Statistics.

de Finetti, B. (1972). *Probability, induction, and statistics.* New York: Wiley.

de Finetti, B. (1974). *Theory of probability*, Vol. 1. New York: Wiley.

de Finetti, B. (1975). *Theory of probability*, Vol. 2. New York: Wiley.

DeGroot, G. J. (2004). *The bomb: A life.* Cambridge, MA: Harvard University Press.

Denis, D. J. (2005). The modern hypothesis testing hybrid: R. A. Fisher's fading influence. *Journal de la Société Française de Statistique, 145*, 5–26.

Efron, B. (1986). Why isn't everyone a Bayesian? *American Statistician, 40*, 1–11.

Falk, R., & Greenbaum, C. W. (1995). Significance tests die hard. *Theory and Psychology, 5*, 396–400.

Ferguson, T. S. (1973). A Bayesian analysis of some nonparametric problems. *Annals of Statistics, 1*, 209–230.

Gelfand, A. E., & Smith, A. F. M. (1990). Sampling based approaches to calculating marginal densities. *Journal of the American Statistical Association, 85*, 398–409.

Gelman, A., Carlin, J. B., Stern, H. S., & Rubin, D. B. (2003). *Bayesian data analysis.* New York: Chapman & Hall.

Gigerenzer, G. (1987). Probabilistic thinking and the fight against subjectivity. In L. Krüger, G. Gigerenzer, & M. Morgan (Eds.), *The probabilistic revolution*, Vol. 2. Cambridge, MA: MIT Press.

Gigerenzer, G. (1993). The superego, the ego, and the id in statistical reasoning. In G. Keren & C. Lewis (Eds.), *A handbook for data analysis in the behavioral sciences: Methodological issues.* Hillsdale, NJ: Lawrence Erlbaum Associates.

Gigerenzer, G. (1998). Surrogates for theories. *Theory and Psychology, 8*, 195–204.

Gigerenzer, G., & Murray, D. J. (1987). *Cognition as intuitive statistics.* Hillsdale, NJ: Lawrence Erlbaum Associates.

Gill, J. (1999). The insignificance of null hypothesis significance testing. *Political Research Quarterly, 52*, 647–674.

Gill, J. (2007). *Bayesian methods for the social and behavioral sciences.* 2nd ed. New York: Chapman & Hall.

Gliner, J. A., Vaske, J. J., & Morgan, G. A. (2004). Null hypothesis significance testing: Effect size matters. *Journal of Human Dimensions of Wildlife Issues, 6*, 291–301.

Grayson, D. A. (1998). The frequentist facade and the flight from evidential inference. *British Journal of Psychology, 89*, 325–345.

Greenwald, A. G. (1975). Consequences of prejudice against the null hypothesis. *Psychological Bulletin, 82*, 1–20.

Greenwald, A. G., Gonzalez, R., Harris, R. J., & Guthrie, D. (1996). Effect sizes and p-values: What should be reported and what should be replicated? *Psychophysiology, 33*, 175–183.

Hager, W. (2000). About some misconceptions and the discontent with statistical tests in psychology. *Methods on Psychological Research, 5*, 1–31.

Howson, C., & Urbach, P. (1993). *Scientific reasoning: The Bayesian approach*, 2nd ed. Chicago: Open Court.

Hunter, J. E. (1997). Needed: A ban on the significance test. *Psychological Science, 8*, 3–7.

Hunter, J. E., & Schmidt, F. L. (1990). *Methods of meta-analysis: Correcting error and bias in research findings*. Beverly Hills, CA: Sage.

Jeffreys, H. (1961). *The theory of probability*. Oxford, UK: Oxford University Press.

Kass, R. E., & Raftery, A. E. (1995). Bayes factors. *Journal of the American Statistical Association, 90*, 773–795.

Kirk, R. E. (1996). Practical significance: A concept whose time has come. *Educational and Psychological Measurement, 56*, 746–759.

Krueger, J. (2001). Null hypothesis significance testing: On the survival of a flawed method. *American Psychologist, 56*, 16–26.

Laplace, P. S. (1774). Mémoire sur la Probabilité des Causes par le Évènemens. *Mémoires de l'Académie Royale des Sciences Presentés par Divers Savans, 6*, 621–656.

Laplace, P. S. (1781). Mémoire sur la Probabilités. *Mémoires de l'Académie Royale des Sciences de Paris, 1778*, 227–332.

Laplace, P. S. (1811). Mémoire sur les Integrales Définies et leur Application aux Probabilités, et Specialemement à Recherche du Milieu Qu'il Faut Chosier Entre les Resultats des Observations. *Mémoires de l'Académie des Sciences de Paris*, 279–347.

Laplace, P. S. (1814). *Essai Philosophique sur les la Probabilités*. Paris: Ve Courcier.

Lecoutre, B., Lecoutre, M. P., & Poitevineau, J. (2001). Uses, abuses and misuses of significance tests in the scientific community: Won't the Bayesian choice be unavoidable? *International Statistical Review, 69*, 399–418.

Lindsay, R. M. (1995). Reconsidering the status of tests of significance: An alternative criterion of adequacy. *Accounting, Organizations and Society, 20*, 35–53.

Little, R. J. A. (2006). Calibrated Bayes: A Bayes/Frequentist roadmap. *American Statistician, 60*, 213–223.

Little, R. J. A. & Rubin, D. B. (1983). On jointly estimating parameters and missing data by maximizing the complete-data likelihood. *The American Statistician, 37*, 218–220.

Little, R. J. A., & Rubin, D. B. (1987). *Statistical analysis with missing data*. New York: Wiley.

Liu, J. S. (1996). Nonparametric hierarchical Bayes via sequential imputations. *Annals of Statistics, 24*, 911–930.

Loftus, G. R. (1991). On the tyranny of hypothesis testing in the social sciences. *Contemporary Psychology, 36*, 102–105.

Loftus, G. R. (1993a). Editorial comment. *Memory and Cognition, 21*, 1–3.

Loftus, G. R. (1993b). Visual data representation and hypothesis testing in the microcomputer age. *Behavior Research Methods, Instrumentation, and Computers, 25*, 250–256.

Loftus, G. R. (1994). The reality of repressed memories. *American Psychologist, 48*, 518–537.

Loftus, G. R. (1996). Psychology will be a much better science when we change the way we analyze data. *Current Directions in Psychological Science, 5*, 161–171.

Loftus, G. R. & Bamber, D. (1990). Weak models, strong models, unidimensional models, and psycho-logical time. *Journal of Experimental Psychology: Learning, Memory, and Cognition, 16*, 916–926.

López, E. (2003) Las pruebas de significación: una polémica abierta. *Bordón, 55*, 241–252.

Macdonald, R. R. (1997). On statistical testing in psychology. *British Journal of Psychology, 88*, 333–349.

McGilchrist, C. A., & Woodyer, K. D. (1975). Note on a distribution-free CUSUM technique. *Technometrics, 17*, 321–325.

Meehl, P. E. (1967). Theory-testing in psychology and physics: A methodological paradox. *Philosophy of Science, 34*, 103–115.

Meehl, P. E. (1978). Theoretical risks and tabular asterisks: Sir Karl, Sir Ronald, and the slow progress of soft psychology. *Journal of Counseling and Clinical Psychology, 46*, 806–834.

Meehl, P. E. (1990). Why summaries of research on psychological theories are often uninterpretable. *Psychological Reports, 66*, 195–244.

Menon, R. (1993). Statistical significance testing should be discontinued in mathematics education research. *Mathematics Education Research Journal, 5*, 4–18.

Morris, C. N. (1983). Parametric empirical Bayes inference: Theory and applications. *Journal of the American Statistical Association, 78*, 47–65.

Neyman, J., & Pearson, E. S. (1928a). On the use and interpretation of certain test criteria for purposes of statistical inference. Part I. *Biometrika, 20A*, 175–240.

Neyman, J., & Pearson, E. S. (1928b). On the use and interpretation of certain test criteria for purposes of statistical inference. Part II. *Biometrika, 20A*, 263–294.

Neyman, J., & Pearson, E. S. (1933a). On the problem of the most efficient test of statistical hypotheses. *Philosophical Transactions of the Royal Statistical Society, Series A, 231*, 289–337.

Neyman, J., & Pearson, E. S. (1933b). The testing of statistical hypotheses in relation to probabilities. *Proceedings of the Cambridge Philosophical Society, 24*, 492–510.

Neyman, J., & Pearson, E. S. (1936a). Contributions to the theory of testings statistical hypotheses. *Statistical Research Memorandum, 1*, 1–37.

Neyman, J., & Pearson, E. S. (1936b). Sufficient statistics and uniformly most powerful tests of statistical hypotheses. *Statistical Research Memorandum, 1*, 113–137.

Nickerson, R. S. (2000). Null hypothesis significance testing: A review of an old and continuing controversy. *Psychological Methods, 5*, 241–301.

Oakes, M. (1986). *Statistical inference: A commentary for the social and behavioral sciences.* New York: Wiley.

Page, E. S. (1957). On problems in which a change in parameter occurs at an unknown point. *Biometrika, 44*, 248–252.

Pettitt, A. N. (1979). A non-parametric approach to the change-point problem. *Applied Statistics, 28*, 126–135.

Pollard, P. (1993). How significant is "significance"? In G. Keren & C. Lewis (Eds.), *A handbook for data analysis in the behavioral sciences: Methodological issues*, Hillsdale, NJ: Lawrence Erlbaum Associates.

Pollard, P., & Richardson, J. T. E. (1987). On the probability of making type one errors. *Psychological Bulletin, 102*, 159–163.

Robinson D. H., & Levin, J. R. (1997). Reflections on statistical and substantive significance, with a slice of replication. *Educational Researcher, 26*, 21–26.

Rosnow, R. L., & Rosenthal, R. (1989). Statistical procedures and the justification of knowledge in psychological science. *American Psychologist, 44*, 1276–84.

Rozeboom, W. W. (1960). The fallacy of the null hypothesis significance test. *Psychological Bulletin, 57*, 416–428.

Rubin, D. B. (1987). *Multiple imputation for nonresponse in surveys*. New York: Wiley.

Schaefer, J. L. (1997). *Analysis of incomplete multivariate data*. London: Chapman & Hall.

Schmidt, F. L. (1996). Statistical significance testing and cumulative knowledge in psychology: Implications for the training of researchers. *Psychological Methods, 1*, 115–129.

Schmidt, F. L., & Hunter, J. E. (1977). Development of a general solution to the problem of validity generalization. *Journal of Applied Psychology, 62*, 529–540.

Sedlmeier, P., & Gigerenzer, G. (1989). Do studies of statistical power have an effect on the power of studies. *Psychological Bulletin, 105*, 309–316.

Sen, A., & Srivastava, M. S. (1975a). On tests for detecting change in mean. *Annals of Statistics, 3*, 98–108.

Sen, A., & Srivastava, M. S. (1975b). Some one-sided tests for changes in level. *Technometrics, 17*, 61–64.

Shaban, S. A. (1980). Change point problem and two-phase regression: An annotated bibliography. *International Statistical Review, 48*, 83–93.

Thompson, B. 2002. What future quantitative social science research could look like: Confidence intervals for effect sizes. *Educational Researcher, 31*, 25–32.

Wald, A. (1950). *Statistical decision functions*. New York: Wiley.

Wilkinson, G. N. (1977). On resolving the controversy in statistical inference. *Journal of the Royal Statistical Society, Series B, 39*, 119–171.

Woodward, W. A., & Gray, H. L. (1993). Global warming and the problem of testing for trend in time-series data. *Journal of Climate, 6*, 953–962.

Yoccoz, N. G. (1991). Use, overuse, and misuse of significance tests in evolutionary biology and ecology. *Bulletin of the Ecological Society of America, 72*, 106–111.

CHAPTER 6

A BOOTSTRAP TEST OF SHAPE INVARIANCE ACROSS DISTRIBUTIONS

Response time (RT), the time taken to complete a task, is a common dependent variable that has been used to draw inferences about the nature of mental processing. While many researchers tend to analyze only mean RT, a growing number are examining entire RT distributions to provide constraint on cognitive and perceptual theories (e.g., Ashby, Tien, & Balakrishnan, 1993; Dzhafarov, 1992; Hockley, 1984; Logan, 1992; Ratcliff, 1978; Ratcliff & Rouder, 1998, 2000; Rouder, 2000; Rouder, Ratcliff, & McKoon, 2000; Theeuwes, 1992, 1994; Townsend & Nozawa, 1995; Van Zandt, Colonius, & Proctor, 2000).

Although there are several methods of analyzing distributions, we advocate that researchers consider how properties of *location*, *scale*, and *shape* change across conditions or populations (Rouder, et al., 2003, 2005). Figure 6.1 provides an example of these properties. The left panel shows the case when only location differs between distributions; the center and right panels show the same for scale and shape, respectively.

Location and scale can be given precise meanings. Let the density of a continuous random variable exist everywhere and be expressed as $f(t \mid \theta_1, \ldots, \theta_p)$, where $\theta_1, \ldots, \theta_p$ are parameters. Let $z = (t - \theta_1)/\theta_2$. We refer to the density as being in location–scale form if there exists a density g such that

$$f(t \mid \theta_1, \ldots, \theta_p) = \frac{1}{\theta_2} g\left(\frac{t - \theta_1}{\theta_2} \mid \theta_3, \ldots, \theta_p\right) \qquad (6.1)$$

If Equation (6.1) holds, θ_1 is referred to as the location parameter and θ_2 as the scale parameter. We will say that θ_3 through θ_p are shape parameters. Many random variables have densities that can be expressed in location–scale form. For the normal, for example, $\theta_1 = \mu$, $\theta_2 = \sigma$, and $g(z) = (2\pi)^{-1}e^{-z^2/2}$. The location parameter corresponds to the mean, the scale parameter corresponds to the standard deviation, and there are no shape parameters. For the exponential with $f(x \mid \tau) = \tau^{-1}e^{-x/\tau}$, $x > 0$, there is no location parameter ($\theta_1 = 0$), $\theta_2 = \tau$, and $g(z) = e^{-z}$. The scale parameter corresponds to τ, and there are no shape or location parameters. Neither the location parameter nor the scale parameter are unique. For example, the location parameter may be the mean, mode, or point at which a distribution first attains mass.

There is an asymmetric relationship between location, scale, and shape parameters and the central moments of a distribution. Changes in location certainly imply changes in mean but not in central moments of higher order than the mean. Changes in scale imply changes in even central moments, including the variance. Changes in scale, in general, may also imply changes in the mean too. For example, the exponential has a single parameter, the rate, whose inverse is a scale parameter. Increasing the rate not only decreases the variance, but decreases the mean as well. Changes in shape, in general, may imply changes in all moments. The left panel shows the case when only location differs between distributions; the center and right panels show the same for scale and shape, respectively.

Figure 6.1   Location–scale–shape taxonomy of differences between distributions. The left, center, and right panels show distributions that differ in location, scale, and shape, respectively.

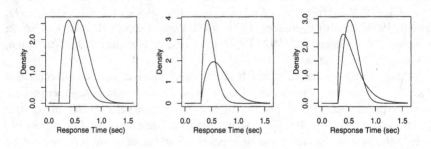

We have advocated the following processing interpretation to these properties. Shape is the most important, as it reflects the underlying mental architecture. Shape changes may be associated with changes in architecture, such as changes in mixtures, adding latent stages, or switching algorithms (e.g., from serial to parallel processing). If shape does change across manipulations, then documenting, exploring, and explaining these changes is surely a good route to better psychological theory. If shape is invariant across manipulations, this invariance licenses the analysis of scale. If shape does not change, and one distribution dominates another, as in the center panel of Figure 6.1, scale is an index of processing speed. It makes little sense to analyze scale across two distributions if they do not share the same shape. For example, it is not meaningful to compare the scale of a normal distribution to that of an exponential distribution. Finally, shift indexes peripheral processes such as traducing the stimulus and executing motor commands (cf. Dhzafarov, 1992; Hsu, 1999; Ratcliff, 1978).

The interpretation above of location, scale, and shape places priority on shape. It makes little sense to compare scale (speeds) of distributions if shape varies. Therefore, it is important to develop methods of assessing shape invariance while allowing location and scale to vary. There are two approaches to the problem: the first is to specify a parametric form with an explicit shape parameter. Examples include the Weibull, log-normal, and inverse Gaussian. We show in the next section that the parametric approach is not robust to misspecification, and therefore is not appropriate. The second approach is nonparametric. The nonparametric approach would be straightforward if the analyst knew the mean and variance of distributions to infinite precision. In this case, the distributions could be shifted and scaled until they had a mean of 0.0 and a variance of 1.0. Then any shape differences would be manifest as differences in the shifted-scaled distributions. Recent developments in testing distribution equalities (e.g., Lombard, 2005; Wilcox, 2006) could then be used to test for shape invariance. Unfortunately, the analyst only knows sample means and variances. The problem is therefore more complex than the problem of assessing distribution equality.

6.1 LACK OF ROBUSTNESS OF A PARAMETRIC TEST

Before developing the nonparametric test, we highlight the problems of a parametric approach by exploring the robustness of a shape test that assumes a Weibull parametric form. The density of a three-parameter Weibull distribution may be given as

$$f(t; \psi, \theta, \beta) = \frac{\beta}{\theta} \left(\frac{t - \psi}{\theta} \right)^{\beta - 1} \exp \left(- \left[\frac{t - \psi}{\theta} \right]^{\beta} \right), \ t > \psi, \ \theta, \beta > 0$$

In this parameterization, parameters ψ, θ, and β serve as the location, scale, and shape of the distribution, respectively. To test the robustness of a Weibull parametric shape test, we performed two simulated experiments, each consisting of 50 participants who each observed 100 trials. In the first experiment, each participant's data came from the Weibull distribution with location, scale, and shape parameter values of .262 s, .199 s, and 1.7, respectively. Because the test assumes Weibull parametric forms, the test is well specified for this experiment. In the second experiment, each participant's data came from an inverse Gaussian distribution. The density of a three-parameter inverse-Gaussian distribution may be given as

$$f(x|\psi, \theta, \beta) = \sqrt{\frac{\theta}{2\pi}} (x - \psi)^{-3/2} \exp \left\{ \frac{-[(x - \psi)\beta - \theta]^2}{2\theta(x - \psi)} \right\}$$

where $x > \psi$ and $\psi, \theta, \beta > 0$. In this parameterization, parameters ψ, θ, and β serve as the location, scale, and shape of the distribution, respectively. Each participant's data had inverse-Gaussian location, scale, and shape parameters of .2 s, 1.2 s, and 5, respectively. Because the test assumes the Weibull form, it is misspecified for these data. The contrast between the two experiments allows for study of the effects of misspecification. The solid and dashed lines in the left panel of Figure 6.2 show the densities of an inverse Gaussian and a Weibull distribution, respectively, with the parameter values given above. These distributions are highly similar, indicating that there is only a small degree of misspecification. Hence, we would hope that the Weibull parametric shape test yields reasonable results, even when applied to inverse Gaussian data.

We fit two Weibull models. The general model consists of individualized location, scale, and shape parameters. Across the 50 participants, therefore, there are $3 \times 50 = 150$ parameters. The restricted model specified that each participant had the same shape; hence, the restricted model had 50 shift parameters, 50 scale parameters, and a single shape parameter (a total of 101 parameters). Whereas "true" shapes did not vary across participants, a well-calibrated test should yield a rate of type I errors close to the nominal value. We estimated the Weibull models with maximum likelihood and assessed the validity of shape invariance with a likelihood ratio test (G^2; Reed & Cressie, 1988; Wilks, 1938). Details of maximization are as follows. For the general model, the negative log-likelihood was minimized for each individual separately. For each individual, there are three parameters, and the three-parameter minimization was performed with the Nedler–Meade simplex algorithm (Nedler & Mead, 1967) in **R** with the `optim()` command. The case is a bit more complicated for the restricted model. We used a nested optimization design. In the inner loop we minimized location and scale for each individual separately for a fixed common shape with the simplex algorithm. Then, in the outer loop, we found the best common shape that minimized negative log-likelihood with the algorithm from Brent (1973) as implemented in **R** with the `optimize()` command. Because each minimization call is done with respect to a small number of parameters, minimization was quick and reliable.

The Weibull model is not regular, and therefore the asymptotic distribution of the test statistic is not guaranteed to follow the chi-square distribution with 49 degrees of freedom ($\text{df}_{\text{general}} - \text{df}_{\text{restricted}} = 150 - 101 = 49$). To assess the effects of irregularity, we simulated the first experiment (Weibull-generated data) 300 times. The cumulative distribution function of the sampling distribution of G^2 across these 300 replicates is shown as line "W" in the right panel of Figure 6.2. The dashed line shows the theoretical chi-square distribution with 49 degrees of freedom. As can be seen, the observed distribution is fairly close to the chi-square distribution even though regularity is violated. The vertical line shows the criterial value for a nominal type I error rate of .05. The observed type I error rate for this nominal value is .083, which represents only a modest inflation.

Figure 6.2   Effects of misspecification on a Weibull parametric test. Left: Weibull (dashed) and inverse-Gaussian (solid) probability density functions are fairly similar. Right: Cumulative distribution functions of the sampling distribution of log-likelihood ratio statistic (G^2) of Weibull test of shape invariance. Lines labeled "W" and "IG" are for Weibull-generated and inverse-Gaussian-generated data, respectively. The dashed line is the theoretical chi-square distribution with 49 df. The vertical line indicates the nominal $\alpha = .05$ criterion; the observed type I error rates are .083 and .853 for the Weibull and inverse-Gaussian-generated data, respectively.

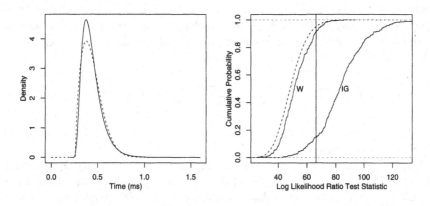

To assess how this seemingly small misspecification of parametric form affected the true sampling distribution of G^2, we performed 300 replications of the second experiment. Line "IG" shows the cumulative distribution function (CDF) of the sampling distribution of G^2. Most of the distribution (85% of the 300 simulation experiments) is above the nominal $\alpha = .05$ criterial value, indicating a massive type I error inflation. Even though the data were generated with shape invariance, the parametric Weibull test is useless, as even small misspecification drive an unreasonably high type I error rate. It is this uselessness that motivates the need for a nonparametric test.

6.2   DEVELOPMENT OF A NONPARAMETRIC SHAPE TEST

We first develop a nonparametric shape test for the case for a single participant who provides data in each of two conditions. Let $\mathbf{x} = x_1, \ldots, x_J$ and $\mathbf{y} = y_1, \ldots, y_J$ denote vectors of J observations in the first and second conditions, respectively. Assume that the data in each condition are independent and identically distributed (e.g., $x_j \overset{iid}{\sim} X$ and $y_j \overset{iid}{\sim} Y$, $j = 1, \ldots, J$). The main question is whether distributions X and Y have the same shape [i.e., whether $(X - \mu_X)/\sigma_X$ and $(Y - \mu_Y)/\sigma_Y$ have the same distribution]. The first stage of the test is the construction of a statistic for shape change; the second stage is the estimation of the sampling distribution of the statistic through resampling.

6.2.0.1 *Step 1: Transform the data* If one knew the population means and standard deviations for distributions X and Y, one could form the standardized samples, say

$$x_j^* = \frac{x_j - \mu_X}{\sigma_X}, \quad y_j^* = \frac{y_j - \mu_Y}{\sigma_Y}, \quad j = 1, \ldots, J \tag{6.2}$$

and perform a nonparametric test for equality of two distributions, such as the Kolmogorov–Smirnov two-sample test (Conover, 1971) or the Cramer–von Mises two-sample test (Anderson, 1962). Since the population means and standard deviations are unknown, it is natural to use the sample means and standard deviations estimated from the data, $\bar{x}$, $\bar{y}$, $\hat{\sigma}_x$, and $\hat{\sigma}_y$, and consider

$$\tilde{x}_j = \frac{x_j - \bar{x}}{\hat{\sigma}_x}, \quad \tilde{y}_j = \frac{y_j - \bar{y}}{\hat{\sigma}_y}, \quad j = 1, \ldots, J \tag{6.3}$$

Our strategy is to use the standardized data in (6.3) in place of the ideal standardized samples in (6.2) in a two-sample test.

Figure 6.3 provides examples of how shape differences in X and Y affect the relationship between $\tilde{X}$ and $\tilde{Y}$. The top-left plot shows two distributions that differ in location and scale, but not in shape. For this example, 500 samples from each of these distributions serve as data. These data were normalized with Equation (6.3); the resulting empirical cumulative distribution functions (ECDFs) are displayed in the top-center plot. As can be seen, these ECDFs are quite similar. The bottom-left panel of Figure 6.3 shows two distributions that differ in location, scale, and shape. Once again, 500 samples from each served as data. The ECDFs of the normalized data are shown in the bottom-center plot. These ECDFs are not as similar as the ECDFs in the top-center panel. When the shapes are the same in data, the normalized distributions are the same; when the shapes are different, the normalized distributions are different. Therefore, to test for shape differences, we test for differences among the normalized data.

6.2.0.2 *Step 2: Quantify differences* One plausible statistic for describing the difference between two sets of normalized samples is a sum-squared difference statistic,

$$d^2 = \sum_{j=1}^{J} (\tilde{x}_{(j)} - \tilde{y}_{(j)})^2 \tag{6.4}$$

where $\tilde{x}_{(j)}$ and $\tilde{y}_{(j)}$ denote the jth-order statistic of the respective normalized sample. If we view $\tilde{x}_{(j)}$ and $\tilde{y}_{(j)}$ as sample quantile estimates for the π_jth quantile with

Figure 6.3 The shape test. Left: Densities underlying data for shape invariance (top row) and shape change (bottom row). Center: Empirical cumulative distribution functions of 500 normalized samples for shape invariance (top row) and shape change (bottom row). Right: Differences in normalized data for shape invariance (top row) and shape change (bottom row).

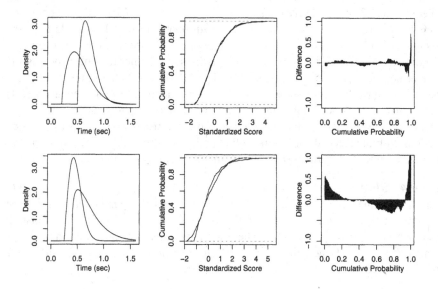

$\pi_j = j/(J+1)$, then $\tilde{x}_{(j)} - \tilde{y}_{(j)}$ is the difference in sample quantiles for the two (normalized) samples. The right panels plot these differences as a function of π_j. If distributions X and Y have the same shape, these segments should be small (top-right panel); as X and Y vary in shape, these segments increase (bottom-right panel). The distance, d^2, is the sum of the squares of these segments.

As an aside, the d^2 statistic is related to the Shapiro–Wilk (1965) one-sample goodness-of-fit test statistic for normality. The Shapiro–Wilk test statistic, denoted r, is the sample correlation coefficient for the order statistics of the sample and the expected value of the order statistics for the standard normal distribution. The sample correlation coefficient is independent of location and scale and can be computed from standardized samples as in Equation (6.3). In our case, if r_{xy} is the sample correlation coefficient for the sorted original data, it is straightforward to show that $d^2 = 2(J-1)(1-r_{xy})$.

The next stage is estimating a sampling distribution for d^2. To our knowledge, this distribution has not appeared in the literature. Thus we implement the test using the bootstrap (Efron, 1979; see Efron & Tibshirani, 1993) to estimate critical values.

6.2.0.3 Step 3: A bootstrap
The bootstrap method is based on resampling from the original data set in some fashion to obtain a "bootstrap" sample. The test statistic is computed for the bootstrap sample. The procedure is repeated many times, and the

empirical distribution of the test statistics computed for the bootstrap samples is used to estimate the true sampling distribution of the test statistic.

Let z be the vector of all the normalized data [i.e., $z = (\tilde{x}, \tilde{y})$]. Under the null hypothesis, the elements of z are approximately independent and identically distributed. [They would be exactly independent if z were computed with equation (6.2) instead of equation (6.3).] The simple bootstrap is based on taking M independent bootstrap samples from z as follows. To obtain the mth bootstrap sample, sample two new vectors of data of length J from z *with replacement*. Suppose that these vectors are denoted $x^{(m)}$ and $y^{(m)}$, respectively. Regardless of whether the null hypothesis is true or not, the bootstrap samples $x^{(m)}$ and $y^{(m)}$ come from the same distribution, hence they have the same shape. Now equations equation (6.3) and equation (6.4) are applied to each bootstrap sample to compute a bootstrap test statistic, which we denote d_m^2. Note that d_m^2 is a test statistic computed from two samples from distributions with the same shape, so the null hypothesis is true. The bootstrap idea is to use these samples to estimate the distribution of d^2 under the null.

6.2.0.4 Step 4: Get the p-value The collection of d_m^2, $m = 1, \ldots, M$, serves to estimate the sampling distribution of d^2 under shape invariance. If d^2 is in the upper tail of the sampling distribution, the null may be rejected. The bootstrap estimate of the p-value is w/M, where w is the number of bootstrap samples for which $d_m^2 > d^2$. Hence, the decision rule for a desired type I error rate of α may be constructed by rejecting the null when $p < \alpha$.

6.3 EXAMPLE

Table 6.1 provides an example of the bootstrap shape test. A single participant provides 10 observations in each of two conditions. The first two rows, labeled x and y, display the data. The rows labeled $\tilde{x}$ and $\tilde{y}$ are normalized in accordance with equation equation (6.3). The distance statistic d^2 is computed in accordance with equation (6.4); for these data, $d^2 = 1.66$.

The rows labeled $x^{(m)}$ and $y^{(m)}$ show a bootstrap step; these values were sampled from the concatenation of $\tilde{x}$ and $\tilde{y}$ with replacement. Note how values from $\tilde{x}$ and $\tilde{y}$ appear in both $x^{(m)}$ and $y^{(m)}$. The following rows show normalized bootstrap samples $\tilde{x}^{(m)}$ and $\tilde{y}^{(m)}$. The d^2 statistic for these two vectors is $d_m^2 = 1.49$. We repeated the process for another 999 bootstrap cycles samples and calculated a squared distance for each. The squared distance from the original sample, $d^2 = 1.79$, is less than 634 of the squared distances from the 1000 bootstrap cycles; hence the p-value for the test is .634. For any reasonable α, the null hypothesis of shape invariance cannot be rejected.

6.4 EXTENSION OF THE SHAPE TEST

It is quite straightforward to extend the test to multiple participants. Consider the case in which each participant provides a set of response times in each condition. Let x_{ij}

Table 6.1 Sample Data, Transformed Data, and Bootstrapped Samples Provide an Example of the Shape Test

x	y	$\tilde{x}$	$\tilde{y}$	$x^{(m)}$	$y^{(m)}$	$\tilde{x}^{(m)}$	$\tilde{y}^{(m)}$
0.073	1.272	-0.928	1.448	0.762	-0.925	1.515	-0.943
0.580	0.903	0.659	0.080	-1.051	-0.515	-1.428	-0.577
0.410	0.524	0.127	-1.326	-0.928	0.127	-1.229	-0.004
0.419	0.827	0.155	-0.202	0.080	-0.515	0.408	-0.577
1.107	0.827	2.308	-0.202	-0.202	0.127	-0.050	-0.004
0.515	1.308	0.455	1.581	0.155	-1.326	0.530	-1.301
0.074	0.835	-0.925	-0.172	0.455	0.659	1.017	0.047
0.233	0.598	-0.427	-1.051	-0.909	1.581	-1.198	1.293
0.205	1.087	-0.515	0.762	-0.202	2.308	-0.050	1.942
0.079	0.634	-0.909	-0.918	0.127	-0.202	0.484	-0.298

and y_{ij} denote the jth response for the ith participant, $i = 1, ..., I$, in the control and treatment conditions, respectively. We assume that each participant has his or her own characteristic shift, scale, and shape in each condition. The hypothesis under consideration is that while this shift and scale may vary across conditions and people, shape only varies across people; it does not vary across conditions. The first step in the bootstrap test for this hypothesis is to sum square distance over participants:

$$d^2 = \sum_i d_i^2,$$

where d_i^2 is the squared distance statistic for each participant. The same summing is applied to each cycle of the bootstrap (i.e., $d_m^2 = \sum_i d_{i,m}^2$, where $d_{i,m}^2$ is the bootstrapped squared distance for the ith person on the mth cycle).

A second extension covers the case in which there are different numbers of observations in the control and treatment conditions. Two modifications are needed. First, the number of samples comprising $x^{(m)}$ and $y^{(m)}$ are adjusted to reflect the numbers of samples in each condition. Second, the distance statistic is modified to account for unequal numbers of samples. Let J^* be the smaller sample size and let $\pi_j = j/(J^* + 1), j = 1, \ldots, J^*$ be an estimate of the CDF of the jth ordered observation for the smaller sample. Next, compute the sequence of π_jth quantiles for $\tilde{x}^{(m)}$ and $\tilde{y}^{(m)}$. These sequences are ordered and the test proceeds as before.

6.5 CHARACTERISTICS OF THE BOOTSTRAP SHAPE TEST

We ran a set of simulations to assess the level (type I error rate) and the power of the bootstrap shape test. The reported simulations consisted of 40 participants each observing 200 trials in each of two conditions. These values denote a moderately large scale experiment in cognitive and perceptual psychology. We repeated this experiment 5000 times with $M = 1000$ resampling cycles. In appplication, we would recommend a much larger value of M, but the current value provides sufficient accuracy to ascertain the approximate level and power of the tests.

The tests were performed across three distributions that have been useful in modeling RT: the Weibull, the ex-Gaussian, and the inverse-Gaussian or Wald. The density of the ex-Gaussian may be given as

$$f(t; \psi, \theta, \beta) = \theta^{-1} \frac{\exp(z\beta^{-1}) + .5\beta^{-2}}{\beta} \Phi(z - \beta^{-1}), \quad \theta, \beta > 0$$

where $z = t - \psi/\theta$ and Φ is the cumulative distribution function of the standard normal. For both of these distributions, ϕ, θ, and *beta* serve as location, scale, and shape parameters, respectively.

The first set of simulations was performed to assess the level or type I error-rate of the shape bootstrap test. In this case, true shape values were held constant across people and conditions. These shape values were chosen such that the skewness of the distributions was approximately .9. The Weibull with this skewness is the middle density in Figure 6.4. In our initial simulations, shifts and scales were varied across each participant-by-item pairing, but this variation had, as expected, no effect on the level and power obtained.

The results of the simulations are shown in Table 6.2. We report the proportion of times the null was rejected at $\alpha = (.01, .05, .10)$. The first three rows of Table 6.2 show the case for shape invariance. For these distributions the bootstrap test is conservative for the Weibull and inverse-Gaussian and well-calibrated for the ex-Gaussian.

Figure 6.4   Weibull distributions with skewnesses of .7, .9, and 1.1 (equated means and variances).

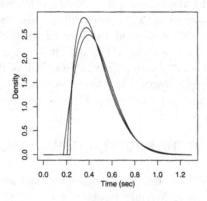

Table 6.2 Proportion of Rejections of the Null Hypothesis Observed in Simulations

Distribution	Shape Parameter Value Cond. 1	Cond. 2.	Type I Error α Level .01	.05	.10	Power α Level .01	.05	.10
			RT Distributions					
Weibull	1.66	1.66	.001	.001	.030	—	—	—
Ex-Gaussian	1.19	1.19	.008	.051	.109	—	—	—
inverse-Gaussian	11.00	11.00	.003	.022	.061	—	—	—
Weibull	1.48	1.90	—	—	—	.957	.992	.998
Ex-Gaussian	1.00	1.43	—	—	—	.440	.725	.836
Inverse-Gaussian	7.40	18.30	—	—	—	.722	.904	.954
			Other Distributions					
Normal	—	—	.001	.004	.015	—	—	—
Logistic	—	—	.005	.040	.098	—	—	—
Exponential	—	—	.015	.080	.159	—	—	—
Pareto	3.00	3.00	.720	.860	.920	—	—	—

We ran a second set of simulations to assess power. To model a large shape change, we chose shape parameter values that correspond to a skewness of .7 and 1.1 in the two conditions, respectively. Weibull densities with these skewnesses are drawn as the outer lines in Figure 6.4. As shown in the Table 6.2, the power is quite high, even when the level is too conservative.

Although the bootstrap shape test is reasonable for the RT distributions explored, the level of the test depends on the distribution. We ran an additional set of simulations with the normal, logistic, exponential, and Pareto distribution with density given as

$$f(t|\psi, \beta) = \beta \frac{\psi^{\beta}}{t^{\beta+1}}, \quad t \geq \psi, \quad \beta > 0$$

As can be seen, the level varies dramatically over these forms. For thin-tailed distributions, such as the normal and Weibull with a shape of 1.7, the bootstrap test is highly conservative. It is much better calibrated for exponential tailed distributions (logistic, ex-Gaussian, inverse-Gaussian, exponential) and ridiculously liberal for fat-tailed distributions (Pareto with a shape of 3.0). Fortunately, RT distribution tails tend to be no fatter than an exponential, as indicated by either flat or rising hazard functions (Burbeck & Luce, 1982; Wolfe, 1998). Therefore, while the bootstrap shape test is appropriate for RT distributions, it may not be so for other domains.

6.6 APPLICATION

To demonstrate the utility of the bootstrap shape test in realistic contexts, we tested empirical data from a priming experiment (Pratte & Rouder, submitted). Participants observed single-digit numbers (2, 3, 4, 6, 7, 8) as targets and had to classify each as being either greater than or less than 5. Prior to target, single-digit primes were displayed for 25 ms and masked. These primes, while barely visible, affected the time it took to classify target. Targets preceded by congruent primes, that is, those primes that elicit the same response as the target, were speeded by 13 ms relative to targets preceded by incongruent primes. In Pratte and Rouder's experiment, 43 participants observed 288 congruent and 288 incongruent trials. Pratte and Rouder excluded responses if (a) they were incorrect (5%), (b) response time was outside a range of 200 ms to 3 s (.3%), or the prime and target were the same digit (so as to preclude repetition priming effects). After exclusions there were an average of 270 incongruent trials and 184 congruent trials per participant.

Data in priming experiments are often displayed as *delta plots* (e.g., de Jong, Liang, & Lauber, 1994; Ridderinkhof, Scheres, Ooserlaan, & Sergeant, 2005). Delta plots are rotated quantile–quantile (QQ) plots; the diagonal of the QQ plot is rotated to the x-axis (Zhang & Kornblum, 1997). To draw a delta plot, RT deciles are computed for each participant in each condition. For each person and each decile, the difference in RT and the average RT between the incongruent and congruent condition are calculated. The difference score is plotted on the y-axis; the average score is plotted on the x-axis. The left panel of Figure 6.5 is the delta plot for this experiment. The 43 gray lines in the center plot show this relationship between difference and average RT. The result is that the priming effect is greatest for the quickest responses and falls off as responses slow. The points show averages at each decile. The y-axis value is the average difference across conditions; the x-axis value is the average of the average across conditions. The decline in priming effect with slower responses is clear; moreover, this decline has previously been observed in near-liminal number priming experiments (Greenwald, Abrams, Naccache, & Dehaene, 2003). It is not clear that the priming effect reverses for the slowest responses. One prevailing interpretation of the decline is that the priming effect is relatively short lived (e.g., Greenwald, Draine, & Abrams, 1996; Neely, 1977). We provide an alternative explanation below after exploring the possibility of shape effects.

The result of a bootstrap shape test is a significant difference in shape across priming conditions ($p < .001$). The right panel of Figure 6.5 provides a supplemental graphical view of the shape test; the ECDF of all 43 participant's p-values is plotted. Under the shape invariance null, this ECDF should be that of a standard uniform and lie on the diagonal. It does not, indicating that values of p_i are less than expected under shape invariance.

The difference in shape across congruent and incongruent conditions is a primary marker of processing. One fruitful avenue to pursue is a mixture explanation. We speculate that participants are consciously aware of the 25-ms primes for some trials and not for others. Indeed, in a separate block, participants were able to classify the prime accurately as greater than or less than 5 on 61% of the trials. Those trials that generate awareness may have a priming effect (e.g., congruent primes that generate awareness result in quicker responses than incongruent primes that generate

Figure 6.5 Result of a bootstrap shape test. Left: Light gray lines show individual delta plots; the solid line with dots is a group average. The late dip below zero is not reliable at the group level. Right: ECDF of p_i. Under the null, the p_i are distributed as a standard uniform (diagonal line). The departure is substantial; shape invariance is rejected.

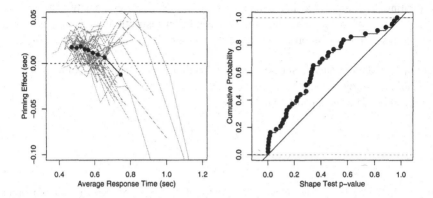

awareness). Primes that do not generate awareness do not generate a congruency effect. This mixture model would produce shape differences. It is also reasonable to speculate that those primes generating awareness happen on trials in which the participants are paying peak attention to the set of events. Such peak attention will lead to relatively fast responses to the target. This explanation accounts for both the decline in delta plots (congruent effects happen when there is peak awareness and short response times) and a shape change.

Our explanation is admittedly post hoc and speculative. The exercise demonstrates how shape testing provides important leads for theoretical development. A shape difference is an indicator to the researcher that there is an important marker of processing that needs further exploration. In this case, we would propose a mixture, but clearly more experimentation is needed. A priming researcher may try to localize the components of the mixture though some manipulation, perhaps through manipulating stimulus-to-target asynchronies (SOAs) or the relative proportion of incongruent to congruent primes.

6.7 CONCLUSION

Understanding the shape of RT distributions is critical for linking models of information processing to data. In this chapter, we show how parametric tests may not be robust to misspecification and provide a nonparametric alternative. The developed test makes use of modern resampling techniques and is relatively easy to implement. The test proves powerful while maintaining adequate control of type I error for RT distributions with tails no fatter than an exponential.

There are two large drawbacks to the current test: First, the calibration of the test depends too greatly on the underlying distribution. It would be desirable to have a test whose level was not so dramatically dependent on the form of the tail. Second, the test applies only to paradigms in which all participants observe stimuli in two conditions. It is not yet evident how to expand the test to factorial designs or continuously measured independent variables. Nonetheless, many experiments in cognition and perception obey the two-condition restriction; in these cases, the bootstrap shape test is appropriate for RT distributions.

Jeffrey N. Rouder, Paul L. Speckman, Douglas Steinley, Michael S. Pratte and Richard D. Morey
University of Missouri, Columbia

REFERENCES

Anderson, T. W. (1962). The choice of the degree of a polynomial regression as a multiple decision problem. *Annals of Mathematical Statistics, 33,* 255–256.

Ashby, F. G., Tien, J.-Y., & Balikrishnan, J. D. (1993). Response time distributions in memory scanning. *Journal of Mathematical Psychology, 37,* 526–555.

Burbeck, S. L., & Luce, R. D. (1982). Evidence from auditory simple reaction times for both change and level detectors. *Perception & Psychophysics, 32,* 117–133.

Conover, W. J. (1971). *Practical nonparametric statistics.* New York: Wiley.

De Jong, R., Liang, C. C., & Lauber, E. (1994). Conditional and unconditional automatiity: A dual-process model of effects of spatial stimulus–response concordance. *Journal of Experimental Psychology: Human Perception and Performance, 20,* 731–750.

Dzhafarov, E. N. (1992). The structure of simple reaction time to step-function signals. *Journal of Mathematical Psychology, 36,* 235–268.

Efron, B. (1979). Bootstrap methods: Another look at the jackknife. *The Annals of Statistics, 7,* 1–26.

Efron, B., & Tibshirani, R. (1993). *An introduction to the bootstrap.* London: Chapman & Hall.

Greenwald, A. G., Abrams, R. L., Naccache, L., & Dehaene, S. (2003). Long-term semantic memory versus contextual memory in unconscious number processing. *Journal of Experimental Psychology: Learning, Memory, & Cognition, 29,* 235–247.

Greenwald, A. G., Draine, S. C., & Abrams, R. L. (1996). Three cognitive markers of unconscious semantic activation. *Science, 273,* 1699–1702.

Hockley, W. E. (1984). Analysis of reaction time distributions in the study of cognitive processes. *Journal of Experimental Psychology: Learning, Memory, & Cognition, 10,* 598–615.

Hsu, Y. F. (1999). Two studies on simple reaction times: I. On the psychophysics of the generalized Pieron's law. II. On estimating minimum detection times using the time estimation paradigm. Unpublished doctoral dissertation, University of California–Irvine.

Logan, G. D. (1992). Shapes of reaction time distributions and shapes of learning curves: A test of the instance theory of automaticity. *Journal of Experimental Psychology: Learning, Memory, & Cognition, 18,* 883–914.

Lombard, F. (2005). Nonparametric confidence bands for a quantile comparison function. *Technometrics, 47,* 364–369.

Neely, J. H. (1977). Semantic priming and retrieval from lexical memory: Roles of inhibitionless spreading activation and limited-capacity attention. *Journal of Experimental Psychology: General, 106,* 226–254.

Pratte, M. S., & Rouder, J. N. (2009). A task-difficulty artifact in subliminal priming. *Attention, Perception, & Psychophysics, 71,* 1276–1283.

Ratcliff, R. (1978). A theory of memory. *Psychological Review, 85,* 59–108.

Ratcliff, R., & Rouder, J. N. (1998). Modeling responses times for decisions between two choices. *Psychological Science, 9,* 347–356.

Ratcliff, R., & Rouder, J. N. (2000). A diffusion model analysis of letter masking. *Journal of Experimental Psychology: Human Perception and Performance, 26,* 127–140.

Reed, T. R. C., & Cressie, N. A. C. (1988). *Goodness-of-fit statistics for discrete multivariate data.* New York: Springer-Verlag.

Ridderinkhof, K. R., Scheres, A., Oosterlaan, J., & Sergeant, J. A. (2005). Delta plots in the study of individual differences: New tools reveal response inhibition deficits in AD/HD that are eliminated by methylphenidate treatment. *Journal of Abnormal Psychology, 114,* 197–215.

Rouder, J. N. (2000). Assessing the roles of change discrimination and luminance integration: Evidence for a hybrid race model of perceptual decision making in luminance discrimination. *Journal of Experimental Psychology: Human Perception and Performance, 26,* 359–378.

Rouder, J. N., Lu, J., Speckman, P. L., Sun, D., & Jiang, Y. (2005). A hierarchical model for estimating response time distributions. *Psychonomic Bulletin and Review, 12,* 195–223.

Rouder, J. N., Ratcliff, R., & McKoon, G. (2000). A neural network model for priming in object recognition. *Psychological Science, 11,* 13–19.

Rouder, J. N., Sun, D., Speckman, P. L., Lu, J., & Zhou, D. (2003). A hierarchical Bayesian statistical framework for response time distributions. *Psychometrika, 68,* 587–604.

Shapiro, S. S., & Wilk, M. B. (1965). An analysis of variance test for normality (complete samples). *Biometrika, 52,* 591–611.

Spieler, D. H., Balota, D. A., & Faust, M. E. (1996). Stroop performance in healthy younger and older adults and in individuals with dementia of the Alzheimer's type. *Journal of Experimental Psychology: Human Perception and Performance, 22,* 461–479.

Theeuwes, J. (1992). Perceptual selectivity for color and form. *Perception & Psychophysics, 51,* 599–606.

Theeuwes, J. (1994). Stimulus-driven capture and attentional set: Selective search for color and visual abrupt onsets. *Journal of Experimental Psychology: Human Perception and Performance, 20,* 799-806.

Townsend, J. T., & Nozawa, G. (1995). On the spatio-temporal properties of elementary perception: An investigation of parallel, serial, and coactive theories. *Journal of Mathematical Psychology, 39,* 321–359.

Van Zandt, T., Colonius, H., & Proctor, R. W. (2000). A comparison of two response time models applied to perceptual matching. *Psychonomic Bulletin and Review, 7*, 208–256.

Wilcox, R. R. (2006). Some results on comparing quantiles of dependent groups. *Communications in Statistics: Simulation & Computation, 35*, 893–900.

Wilks, S. S. (1938). The large-sample distribution of the likelihood ratio for testing composite hypotheses. *Annals of Mathematical Statistics, 9*, 60–62.

Wolfe, J. M. (1998). What can 1 million trials tell us about visual search? *Psychological Science, 9*, 33–39.

Zhang, J., & Kornblum, S. (1997). Distributional analyses and De Jong, Liang, and Lauber's (1994) dual-process model of the Simon effect. *Journal of Experimental Psychology: Human Perception and Performance, 23*, 1543–1551.

CHAPTER 7

STATISTICAL SOFTWARE FOR THE SOCIAL SCIENCES

Contemporary social science research and evaluation require access to a statistical package that has the data management capabilities, graphical support, and statistical functions and procedures necessary to perform required analyses. In fact, many research endeavors necessitate the use of more than one statistical package, as well as several ancillary utility packages. In this chapter, we describe the most popular software applications used in social science research as well as a few that probably should get more use.

In the 10 years I have served as the software reviews editor for *The American Statistician*, a publication of the *American Statistical Association*, I have had the opportunity to become acquainted with a large number of statistical software packages and support utilities. In the latter category I place software such as Stat/Transfer, which is used to convert file formats between nearly every major statistical application, ePrint 5 Professional, a low-cost program used to convert any Windows format to a PDF image, and nQuery Advisor, used to access appropriate sample size and power for a given research project. These types of programs typically provide the user with a single capability, which they do well. Of course, some are better than others, and some may have more capability, but not enough to warrant the excess cost compared with similar packages.

In any case, I have found that several heavily advertised and well-used applications may not in fact provide the research scientist with the statistical capabilities needed to evaluate their data properly, at least not as well as perhaps some lesser known packages. I hope to provide the reader with some of the insights I have gained over the years in determining which package may be most appropriate for particular needs.

I first outline various statistical capabilities that an appropriate software application should have for proper social science analysis. Subsequently, I provide an overview of the major statistical software applications with the view of how they meet the requirements detailed for social science research. Of course, an analyst may need only one statistical procedure when working on a particular project. In this case they may find that several of the applications listed suit their specific requirements.

7.1 SOCIAL SCIENCE RESEARCH: PRIMARY CAPABILITIES

- Data management

    Data transformations, appending, merging, sorting, labeling, file conversion, import and export, data verification, row and column management, subset defining

- Graphical support

    Line charts; area plots; histograms; scatterplots; bar charts; pie charts; hi-lo charts; regression diagnostic graphs; spike plots; stem-and-leaf plots; survival plots; nonparametric smoothers; distribution Q–Q plots; other distribution diagnostic plots; ROC graphics

- Descriptive statistics

    Summary statistics, cross-tabulations, correlations, t-tests, proportion tests, association statistics, equality-of-variance tests, confidence intervals, nonparametric tests, normality tests, random number generation, equality of variance tests

- ANOVA-type DOE

    One-way, two-way, n-way ANOVA and analysis of covariance, randomized block designs, contrasts, factorial, unbalanced, mixed, and nested designs, repeated measures

- Regression modeling

    Continuous response models: OLS; lognormal; two- and three-stage least squares; polynomial; fractional polynomial

    Generalized linear models: Gaussian; gamma; inverse Gaussian; binomial (logit, probit, clog-log, log-log); Poisson; geometric; negative binomial; power

Binary response models: logit; probit; clog-log

Proportional of grouped response models: logit; probit clog-log

Count response models: Poisson and Poisson with rates; negative binomial (NB1 and NB2); zero-inflated models; truncated; censored, hurdle models

Ordered response models: multinomial logit, probit

Unordered response models: ordered binomial; generalized ordered binomial models

Panel models: unconditional and conditional fixed effects; random effects; generalized estimating equations

Mixed linear models: random intercept; random coefficient of parameter; hierarchical models; spatial analysis

Nonparametric models: generalized additive models; smoothed regressions; quantile regression

Nonlinear models

- Multivariate models

  Multivariate analysis of variance

- Survival or event history models

  Kaplan–Meier; life tables; Cox proportional hazards models, with time-varying covariates; parametric models; Mantel–Haenszel tests; discrete response models

- Classification models

  Discriminant analysis, cluster analysis, PCA, factor analysis, correspondence analysis, multidimensional scaling; classification trees

- Nonparametric tests

  Wilcoxon–Mann–Whitney and Wilcoxon signed rank tests; logrank tests; Kruskal–Wallis tests; Spearman and Kendall correlations; Kolmogorov–Smirnov tests; exact binomial CIs

- Resampling statistics

  Bootstrap; jackknife; permutation tests; Monte Carlo simulation

- Exact statistical analysis

  For small and/or unbalanced tables; logistic and Poisson regression models having few observations, or where covariates are highly unbalanced

- Programming language

Batch file processing; statistical procedure creation; matrix programming, function creation; low-level graphic programming

- Other capabilities relevant for most social science research

  Survey methods; time-series methods; quality control; six sigma

- Utilities useful for analysis

  Sample-size analysis; built-in spreadsheet; built-in text editor for program creation; built-in Internet accessibility; Web download of updates; technical support; permanent license; free technical and user support

7.2 STATISTICAL SOCIAL SCIENCE STATISTICAL APPLICATIONS

The following statistical packages are those most used in applications in the social sciences, not counting econometrics. If we include econometrics, we must add LIMDEP to the following list: *SAS, SPSS, Stata, R, Statistics, StatXact/LogXact.* Useful utilities include: *Stat/Transfer, ePrint Professional*, and *nQuery Advisor*.

7.2.1 R

R is a freeware package. R is at present the most commonly used software application among academicians at the larger universities. R is based on a programming language similar to that of S and the commercial S-Plus and has become extremely popular both in the United States and Canada as well as abroad. Since the application is dependent on user-created statistical procedures and functions, the number of its statistical capabilities grows each month.

A major drawback with R is that there is no error-catching system other than peer review. There exists a group of R experts who devote considerable time to assuring appropriate user interface as well as evaluating primary functionality, but their work is purely gratis. Although overseeing cadres of statisticians have thus far worked hard to guarantee the accuracy of the foremost R procedures, they still cannot evaluate all of the libraries of programs that have been attached to the overall application. That is, the user should employ caution when analyzing or modeling data with newly created libraries or programs.

On the other hand, procedures that have been with the package for some time have probably been evaluated by peers. *The Journal of Statistical Software*, of which I am an editor, evaluates submissions of R software to the archives. This helps assure the quality of the software, but again, not all prepared procedures are given to review.

R provides the user with a vast range of statistical capabilities. Nearly all submissions to the R archives reflect the interests of the creator. I suspect that programs have been designed for nearly every traditional descriptive statistic in the literature. Moreover, few procedures or functions found in commercial software are not in R. There are a few, to be sure, but not many. At present I know of no R program for the canonical negative binomial, nor for censored survival count models. There is little doubt that such R capabilities will be forthcoming, but it does indicate that R is not a finished product.

It appears to me that most of the libraries submitted to R deal with some type of modeling procedure, or with goodness-of-fit and model testing concerns. This is certainly advantageous for those engaged in social science research. Many of our research endeavors relate in some manner to the modeling of data. R provides the social statistician with an easy-to-use, inexpensive, tool by which to understand and graph the data at hand.

It appears that R will be a well-used statistical application for many years to come. Many universities require their graduate statistics students, as well as those taking graduate statistics courses in the social sciences, to learn R. In fact, universities such as UCLA and Stanford University make R competence mandatory for obtaining a graduate-level degree in statistically related disciplines.

See http://www.r-project.org/ for instructions on downloading and other document information. R software may be downloaded from a number of mirror sites from around the world. Some are faster than others. If you find that downloading the software is taking an inordinate amount of time, I suggest breaking out of the download process and trying another site. Generally speaking, the closer the mirror site, the faster the download. The only caveat is the download speed that is listed with the mirror. Be sure that you have the maximum download speed compatible with your computer.

7.2.2 SAS

SAS is perhaps the father of statistical and data management systems. For the past 30 years it has been the predominant statistical application in the business and pharma-ceutical industries. Some of this dominance has dwindled in the past 10 years with the rise in popularity of packages such as Stata and, of course, R. But it is still the *franca lingua* of the big business world.

SAS wants to foster the notion that it is not just a statistical application but, rather, an integrated statistics–data management–graphics system that is to be used as the only business tool at a site. As such, it is has rather extensive capabilities, having few major omissions. The downside is its extremely high cost and the difficulty in mastering SAS programming. In fact, at the majority of large businesses using SAS, statisticians and SAS programmers are hired into and work in separate departments. SAS programmers are an entity unto themselves.

SAS is rarely purchased by individuals, due to its high cost. However, academics are typically able to use SAS via a site license through their university bookstore or computing center. For an average cost of some $120, faculty members may obtain a license to use SAS on their personal computer, but they typically have access to only the previous-to-current version. Many site license users are therefore precluded from using SAS's newest offerings.

A license for use is typically for one year only. After a year, the user is required to pay an update fee for each software module acquired. New options generally required payment of full new-user fees. Technical support is not free; one must purchase access to maintenance support. Requests for support are submitted via the Web or phone, with a response forthcoming within a few days. Other more responsive plans are available, but only for a substantially higher fee.

If one does not have access to a site license, the software will impose a considerable setback on one's checkbook. The software must be renewed each year, with a cost attached to each unit acquired. The first-year license for the modules required to deal with most types of decent social science research can cost well in excess of $5000. SAS has numerous fee plans; it is therefore not possible to cite an absolute license cost that hold for all situations. The bottom line, though, is that it is an expensive application. For large corporate entities, I have heard of yearly SAS charges approximating $1,000,000.

SAS comes with a host of technical support books and booklets. Annual SUGI conferences are held where users can discuss problems among themselves and obtain hints from SAS experts and from SAS staff themselves. I have myself presented a paper at one of the SUGI meetings. There are literally hundreds of papers shared, and workshops given, to assist users with SAS programming and statistical issues.

Many textbooks discuss the particular subject of concern using SAS for examples. Together with this support, this all, of course, creates an atmosphere amenable to the SAS user.

On the downside I have observed that SAS seems to be losing its user base in areas such as social science research and in health outcomes analysis. Journal articles on straightforward biostatistical analysis also seem to be leaning away from SAS. As we shall find shortly, Stata has taken a substantial share of statistical research from SAS.

7.2.3   SPSS

SPSS (http://www.spss.com/) has been a mainstay among social scientists for over 30 years. Although found in the business community, it is not nearly as prevalent as is SAS. However, it does have a good product share. SPSS, like SAS, is highly modularized. One purchases the basic package, consisting of data management and graphical and basic statistical capabilities. Advanced statistical modules, including more sophisticated statistical modeling tools such as missing values analysis, exact statistics, time series, classification trees, categorical data analysis, and so on, are available for an extra charge. The charge to a commercial entity for the advanced statistics module, which includes programs for generalized linear models, generalized estimating equations, and generalized linear mixed models, is $899. If already part of your package, an update is $125. The regression package, including binary logistic regression, multinomial logistic regression, probit analysis, and nonlinear regression, is an additional $899. The base package is $1599. Maintenance (technical support) costs an additional $400. If the user desires a well-rounded statistical package, the cost will come to some $6000 to $8000. Yearly updates will be somewhere in the range of $2000. It is not an inexpensive application.

SPSS has since its inception had major shortcomings in particular areas of statistics. Some of these areas have been extremely important, and have resulted, in my opinion, in SPSS losing thousands of users. For example, until the current 15.0 version, SPSS had no Poisson regression procedure. In fact, it had no standard regression model for counts (e.g., Poisson, negative binomial, geometric, ZIP/ZINB). The makeshift workaround provided by SPSS technical support for a simple Poisson regression model was simply tortuous. Aside from the complexity, one could not have a combination of continuous and binary predictors; and zero counts were not allowed. The workaround was for all purposes useless. Both the Poisson and negative binomial models were made part of version 15 after continuous criticisms that I made in my reviews of SPSS.

Moreover, SPSS was one of the last major software packages to provide its users with the ability to model using generalized linear models (GLM). Generalized linear models have been part of SAS, Stata, S-Plus (and R), GENSTAT, and several other major software applications since at least the early 1990s. It is a covering algorithm allowing the user to select an exponential family-based regression model by simply choosing a family and link function. The Gaussian family, with the canonical identity link, where the linear predictor, xb, is identical to the fitted value, m or y is the same model as OLS regression. Selecting the binomial family and logit link from the GLM menu, provides the user with a logistic regression. A Poisson family and log link is a Poisson regression.

There were several other major procedures unavailable to SPSS users prior to version 15; however, SPSS has made a substantial effort to respond to user requests. Procedures such as generalized linear models, generalized estimating equations, Poisson regression, and several other important procedures are now available to SPSS users. To be sure, there are other important procedures of interest to social scientists still unavailable to SPSS users. If SPSS is as responsive to user requests for version 16 as they were in version 15, SPSS will be nearly as competitive as SAS and STATA.

SPSS now has a wide variety of statistical, data management, and graphical capabilities. It is suitable for most social scientific projects. Faculty can purchase a site license for SPSS through their university bookstore or computing center for about the same cost as SAS. However, not all universities have entered into a site license agreement with SAS or SPSS. If not purchased through a site license, SPSS is quite expensive, but not nearly so much as SAS (see Hilbe, 2005a).

7.2.4 Stata

Stata is over 20 years old, but did not start to take off until 1991, with the initial issue of the *Stata Technical Bulletin* (STB). The STB was a user-based journal in which Stata users published new statistical procedures, novel data management capabilities, functions and random number generators, and a host of entries useful to the Stata community. The capability of the software skyrocketed such that in five years it was a leading statistical application worldwide. It did not hurt that Stata was adopted by the Health Care Financing Administration (HCFA), which regulates Medicare, as the official software for Medicare patient cost and care outcomes analysis. Health care statisticians wrote many new procedures as well, making it one of the most comprehensive statistical applications on the market. In fact, it currently has a greater range of regression and modeling procedures than any other software. For a thorough review of the package, read Hilbe (2005b).

The key to Stata is its programming language. The higher language programming facility provides users with an easy way to expand Stata's statistical offerings and provides an easy way for users to shape the application to their own liking. It has a comprehensive menu-based system which can be used to find the most appropriate model to employ with a particular data situation. Each command has its own help file, which can be requested by the menu by typing at the command line:

```
help <command:>
```

For example,

```
. help poisson
```

will help on the Poisson regression procedure.

Stata comes with free technical support as well as a vast cadre of users who regularly communicate with one another over the Stata Listserver, located at Harvard University. If one does not know how to use a particular application, or has a question regarding statistical methodology in general, an answer can usually be found to a query placed on the server within a day. An 800 number provides immediate—and free—contact to Stata's support.

Stata Corp, located in College Station, Texas, also owns Stata Press, which has published some 10 full-fledged books dealing with such subjects as *Generalized Linear Models and Extensions, Multilevel and Longitudinal Modeling Using Stata, An Introduction to Survival Analysis Using Stata, An Introduction to Modern Econometrics Using Stata, Regression Models for Categorical Dependent Variables Using Stata*, and five others, several of which are in their second and third editions. Five more are scheduled for publication in 2007.

Stata Corp also provides users with Web-based courses on programming, survival analysis, and other areas of interest. These are just some of the resources Stata users have to learn both statistics in general and how to engage in statistics using Stata. In a sense, Stata is like R in that users have authored numerous procedures and functions that are now part of the official package. The difference, of course, is on the one hand, the widespread technical support offered by Stata Corp and other Stata users, and on the other hand, the cost compared with R. R is free; Stata costs $895 for the top-ranked educational single-user package (SE/ 9), which includes 15 reference and application manuals, much of which is already on the Web site. The most expensive single computer commercial application is $1595, a truly small cost for what one gets.

Readers should go to www.stata.com and www.stata-press.com for more information about the product, as well as Hilbe (2005b). The latter is a comprehensive overview of the Stata 9 package and includes multiple examples of use and programming: for example, matrix program code, maximum likelihood program code, iteratively re-weighted least squares (IRLS)–based program code, nonlinear program code, and function program code.

7.2.5 STATISTICA

STATISTICA is primarily a multivariate modeling and data mining application, but also has the ability to provide nearly all basic statistical procedures. Currently in version 7.0, the basic STATISTICA costs $1195, which includes two printed reference books and 50 $M\Omega$ of compressed electronic documentation. The separate STATISTICA Data Miner package is quite expensive at $15,000, with a 20% yearly maintenance license fee. Computing laboratory centers at large universities, research institutes, and health care outcomes centers are the foremost Data Miner users.

The basic STATISTICA 7 package provides a vast number of statistical procedures, functions, and data management capabilities. Some of the procedures, however, do not have the scope or functionality for in-depth analysis that an application such as Stata or SAS may have. On the other hand, these two packages provide the user with nearly every option conceived of for many of their respective procedures; few software applications have their depth of options. For example, Stata and SAS provide the ability for those modeling a logistic regression to employ bootstrapped, jackknifed, robust, or even Newy–West standard errors to its parameter estimates. All of the analyses detailed in Hosmer and Lemeshow's *Applied Logistic Regression*, perhaps the most noted text in the area, are viable Stata options. But because STATISTICA does not have as many options as Stata or SAS does not take away from its use or value in social science research.

To reiterate, STATISTICA has a full range of standard statistical package procedures, functions, graphics, and data management tools. It emphasizes multivariate statistics, but offers a comprehensive design-of-experiment (DOE) facility, with a clear industrial application bent. It also offers a very nice range of ANOVA designs, comparable to those of SAS. Complementing STATISTICA's DOE capability, the software provides truly comprehensive statistical support for industrial and business uses. In fact, it is an excellent business application, with obvious value to social scientists.

The STATISTICA Data Miner provides the following major capabilities: independent components analysis, support vector machines, k-nearest neighbor cluster analytic methods, stochastic gradient boosted tree analysis, EM clustering, multivariate adaptive regression splines, classification and regression trees, CHAID trees, neural networks, intelligent problem solvers, generalized additive models, a general slicer/dicer and drill/down explorer, plus all the features of the basic STATISTICA package. Without a doubt, the STATISTICA Data Miner is the most complete data mining application on the market. The graphical support accompanying the data mining facility is unequaled. The only prohibitive feature is the cost, especially for small work units or for individuals.

Mention should be made of the excellent reference manual, authored independently by Thomas Hill and Pawel Lewicki, both of Tulsa University. The authors are recognized authorities in the area of Data Mining and multivariate methods, which readily comes through when reading how examples are evaluated. The text is a valuable resource in itself, independent of the software.

I recommend reading Hilbe's overview (Hilbe, 2007a) for a review of the major capabilities and limitations of both the base package and the Data Miner.

7.2.6 StatXact/LogXact

As of this writing, version 7 is the current version of both StatXact and LogXact. However, version 8 of both applications will be released in the near future. My remarks relate to the final beta version of the software, with the expectation that no changes will be made in the interim.

StatXact is software with a specific purpose. Concerned primarily with the analysis of contingency tables, the software calculates the highly iterative permutations that result in exact nonparametric p-values for F, χ^2, and related statistics. In addition, exact nonparametric p-values are produced for nearly every conceivable test statistic, including, for example, logrank tests, generalized Wilcoxon–Gehan tests, Kolmogorov–Smirov tests, runs and signs tests, K-sample median tests, two related and two independent binomial samples tests, general permutation tests, and tests of correlated samples. The user is provided with literally hundreds of permutational inference tests.

The StatXact reference manual reads like a textbook on exact statistical methods. There is even a section taking the reader step by step through an analysis concluding with an exact p-value. In fact, I used the reference manual as a text for a course I taught in exact statistical methods.

A drawback of using highly iterative techniques is the computer itself. That is, the computer hosting the analysis may not have sufficient residual RAM to run the routine. In such cases, the software drops out of the algorithm for calculating exact statistics and uses a Monte Carlo method instead. Monte Carlo methods are sometimes confused with bootstrap. They are not the same. Monte Carlo samples are taken from the true permutation distribution of the test statistic, resulting in an accurate estimate of the exact p-value. In Monte Carlo simulation, samples are taken from an assumed distribution. Bootstrapping, on the other hand, resamples population data in order to obtain an empirical sampling distribution of the parameter estimates. The important point here is that Monte Carlo methods provide p-values that are typically very close to the true exact values. Therefore, if the data set is too large for the computer's RAM, Monte Carlo algorithms provide accurate p-values, at least more accurate than can be obtained by using standard asymptotic methods.

StatXact provides output on all three p-values, if possible: exact, Monte Carlo, and asymptotic. Unfortunately, though, StatXact does not provide the ability to calculate exact parametric p-values (e.g., t-tests). Only one package does such calculations: XPro. This software limits itself to the calculation of basic OLS regression, ANOVA, t-tests, and a few other parametric models. Unfortunately XPro's interface is poor, and it acts only on text data set up in a specific manner. XPro resembles state-of-the-art 1995 software. Researchers would be benefited if the manufacturerers of XPro would update the interface, or if Cytel, the manufacturers of StatXact, would integrate parametric methods into its application.

This brings us to LogXact, the Cytel application that provides researchers with exact p-values for logistic and Poisson regression parameter estimates. When the data to be analyzed are small or highly unbalanced, traditional asymptotic methods fail. In addition, if there is perfect prediction, of if all observations having the same covariate pattern have the same response (i.e, all one or all zero), asymptotic methods will fail. The algorithm will simply not converge, or if it does, the parameter estimates yield exorbitant values. Monte Carlo p-values are given on output if the data are such that exact statistics cannot be calculated.

Logistic regression can be parameterized with the response taking only binary 1/0 values, or by having the response take the form of a proportion. That is, a grouped or proportional logistic regression may have a response in which a numerator contains the number of times a covariate pattern has a response of 1, with the denominator containing the value representing the number of covariate patterns with an identical configuration. If a binary response logistic regression model has no continuous predictors, converting a binary response model to a grouped model is comparatively simple. There are very good reasons why a researcher may want to do this, but this takes us beyond the present discussion. In any case, STATISTICA does not have the capability to model grouped logistic models, and SPSS gained the capability only through its new GLZ (its name for the generalized linear models command). Interested readers should access Hardin and Hilbe (2007) for a complete discussion.

LogXact also provides the user with the capability to model Poisson regression, with or without rates. Neither SPSS nor STATISTICA allows for a rate parameterization; SAS and Stata do. In addition, users can employ LogXact to engage in a variety of standard asymptotic statistical modeling tasks. Included are probit and complementary loglog regression, polytomous response regression models, and the ability to appropriately adjust models with missing values.

StatXact and LogXact products come with a covering module called Cytel Studio. Aside from providing extensive data management and graphing capabilities, Studio allows users to engage in all of the major types of univariate modeling, as well as the following multivariate models: canonical correlation analysis, discriminant analysis, PCA and factor analysis, and MANOVA. It also gives the user a complete sample size analysis program, thus saving the cost of purchasing a stand-a-lone sample size application; such as nQuery Advisor.

The combination of the two applications, plus Cytel Studio, ranks the product in the realm of full-fledged statistical packages; it is no longer a niche product. The fact that it is so comprehensive warrants its licensing fees:

	StatXact	LogXact
Commercial	$1400	$860
Academic	$995	$510
Yearly support	$350	$240

The Cytel web site is www.cytel.com

7.3 STATISTICAL APPLICATION UTILITIES

I next provide an overview of three utilities, each of which has proved useful to the social science researcher. The first is a file conversion package, the second primarily a package to convert any Windows application to PDF format, and the last, nQuery Advisor, a sample-size utility.

7.3.1 Stat/Transfer

Stat/Transfer, by Circle Systems in Seattle, is a low-cost file conversion program that allows researchers to convert one file to another. For example, most of my corporate clients give me data in the form of an Excel spreadsheet. Less frequently, they present it to me in Access. Rather than use the extremely limited statistical capabilities offered by Excel or Access, I convert the data to the statistical application I believe to be most suitable for analysis. If the data are in the form of a table, with relatively few counts in specific cells, I'll probably convert it for use with StatXact. If the data consist of a large number of observations, I will probably convert data from Excel or Access to Stata. Finally, if huge numbers of data elements are involved, and I want to search the data for trends or to employ various classification analyses, then I'll probably turn to STASTISTICA's Data Miner. The conversion of choice is Stat/Transfer.

Stat/Transfer converts observations, variables, labels, definitions, and so on, between older as well as the current versions of: Lotus 123, Access, ASCII text, dBase, EpiInfo, Excel, FoxPro, Gauss, JMP, LIMDEP, MatLab, Mineset, Minitab, Nlogit, ODBC Data Source, Osiris, Paradox, Quattro Pro, R Workspace, SAS (all varieties of format), S-Plus, SPSS data file and portable, Stata, STATISTICA, and Systat.

Stat/Transfer has impressive data management capabilities. Variables can be transformed to a wide variety of forms, merging, matching, and observation/variable selection are allowed, variable labels can be assigned, and variables may have their formats changed. Selection of files to be converted, as well as the many options, are easily manipulated—quite unlike some of Stat/Transfer's competitors (e.g., DBMS/COPY).

Stat/Transfer version 9.0, which came out in April 2007, comes with a corporate cost of $295 and an academic rate of $179. Upgrade charges are comparatively minimal.

The application is available to try out. If one wishes to have it on a permanent basis, a license key is provided via the Web after payment of the cost. Readers can learn more about the package by going to: http://www.stattransfer.com/.

7.3.2 ePrint Professional

Currently in version 5, ePrint Professional, by Leadtools, is an expensive utility (i.e., $49) that converts between Windows files. In addition, the application provides an easy mechanism to convert a Windows file to a PDF image. The image can be watermarked in a manner specified by the user and can have embedded URLs to allow easy access to references. Therefore, if a teacher constructs an MS Word document for use with a class, the document may be converted to a PDF image together with URLs. HTML pages can also be designed, and converted from Word format.

Although the vendor claims that the application can easily convert between any Windows applications, when one uses it to convert a PDF document to Word or Excel format, the results may not be what the user expects. Each line in a Word document is encased in a box for formatting purposes. One cannot edit such a document or change it in any manner from how it appeared in PDF image form. One may de-select formatting boxes, but the resulting document can then take any form, so it is not recommended.

For the cost involved, I use the package strictly for the creation of PDF documents. In this regard, ePrint has essentially the same capabilities as those of Adobe but at substantially less cost.

PDF images are easy ways to protect a document from change and reformatting when transferred over the Web. Document ownership is easily maintained as well.

ePrint is compatible with all Windows platforms, including Vista. More information can be obtained by going to the vendor web site: http://www.eprintdriver.com/. Additionally, those interested can review Hilbe (2007b) for a more detailed evaluation of the package.

7.3.3 nQuery Advisor

nQuery Advisor, by Statistical Solutions, has been in existence for many years. Without a doubt, it is the premiere stand-a-lone sample-size application on the market. Other applications, including SAS, SPSS, Stata, STATISTICA, and StatXact have excellent sample-size calculation capabilities, but nQuery Advisor provides the user with graphical representations of differential p-value and power values for a range of specified observations or cases, or any combination of the above. nQuery Advisor also provides tutorials to assist the user in deciding between alternatives. All in all it is a fine and reliable sample-size package.

nQuery Advisor version 6 is not inexpensive. The commercial price of the package is $1195, with a hardcopy of the reference manual costing an additional $100. An upgrade from a previously licensed version 5.0 is $595, rather steep for an upgrade price. The academic price is $750, a substantial reduction. The reference manual still costs $100, but the upgrade from version 5 is $495.

If a researcher is using one of the general-purpose applications that have been discussed in this chapter, they will be able to use the built-in sample-size calculation algorithms that come with the respective package. This should, in practice, be all that is required, but if one desires the advice and graphics, and believes the extra cost to be worth it, nQuery Advisor is a fine utility package.

Readers can obtain more information about the application by accessing the vendors web site at

```
http://www.statsol.ie/html/nquery/nquery_home.html
```

7.4 SUMMARY COMMENTS

Social scientists have requirements that are in many respects different from those in many other disciplines. They typically deal with data having large numbers of observations and variables. In fact, there is perhaps no other discipline that handles more variables than in social science research. This is due to the fact that the majority of social surveys include items on a number of issues.

This brings us to another point. Social scientists often deal specifically with survey data. Analysis of survey data requires that the software being used for analysis has the capability of employing probability or sample weights and allows the user to create stratified designs, including stratified multiple-stage designs with primary, secondary, and lower-stage sampling units. Of course, survey analysis sometimes requires fixed and random effects designs as well. In addition, poststratification methods which provide for more efficient variance estimates are vital to social science research. At the very least, software used for social science research must provide a variety of methods for adjusting the standard errors of parameter estimates, including jackknifing, bootstrapping, and implementation of Newey–West and other robust variance estimators. Stratification techniques must be easily applied to data, together with defining of the appropriate adjustments. Few software packages provide these capabilities.

SAS and the survey modules designed for SPSS have a wide range of survey statistics. Stata has a comprehensive repertoire of procedures and adjustments for survey modeling, and R is gaining new methods each year. Even LogXact incorporates a number of survey adjustments into its capabilities. STATISTICA does not support detailed survey analysis, but, as has been discussed, does provide the means to prune data to obtain the most meaningful set of predictors for a given study.

Unfortunately, some of the more popular statistical packages on the market fail to incorporate needed survey methods. In my opinion, S-Plus, SYSTAT, Minitab, StatGraphics, JMP, NCSS base, and similar packages probably should not be used for serious social science research. S-Plus has a variety of user-created commands that are used for such research, but they are not built-in as part of the application.

I will end here with my recommendations. However, they are simply recommendations, based on my own observations and predilections gained through use as well as in my role as software reviews editor for *The American Statistician* since 1997.

Like a long-used toolbox, a statistical application or utility program may have a personal element attached. If an application was used throughout graduate school, or if the application has been used for many years with good success, it may be difficult to convert to another package, even if one's current software does not provide the tools necessary to engage in appropriate analysis. We tend to acquiesce to that to which we are accustomed. In addition, learning curves for all applications, with the exception of utilities, are at times quite steep. Few have the time or desire to devote learning the in's and out's a new application. However, at times it must be done if we are to maintain statistical credibility.

For my own work in social science and health outcomes research, I have found that Stata is superior to any other general purpose statistical package. I again refer the reader to my review (Hilbe, 2005b). Most statisticians with whom I work or with whom I am associated professionally, use Stata as their primary statistical application. This is not to say, however, that other packages are not used for special purposes. I use StatXact for nearly all cross-tabulation analysis and LogXact when modeling small or unbalanced data. Even though I did not discuss it in this chapter, I have used LIMDEP, a leading econometric application, when modeling certain types of complex count response models (e.g., negative binomial random coefficient mixed models). But for all else I use Stata. If it does not have the built-in model I desire, I usually program it myself using Stata's maximum likelihood and matrix programming capabilities. There are very few procedures, however, that cannot be found in Stata or that cannot be downloaded from user sites. However, I must reiterate that my conclusions may not be shared by all.

The important message in this chapter is that appropriate social science research necessitates the use of software that maximizes our understanding of the data. One must employ methods that not only provide, for example, a table of parameter estimates and confidence intervals, but also allow the researcher to evaluate the goodness of fit of a model and compare it to other possible models. When modeling, the appropriate software should allow the user to adjust standard errors if necessary, to evaluate the model using an AIC or BIC statistic, to provide likelihood ratio tests, to provide a complement of residuals appropriate to the model, and so on. This is what makes modeling an art and makes modeling a challenge. If ones software does not have these capabilities, it should be exchanged for one that does, regardless of the cost of the learning curve involved. It is the price of being a research statistician.

Joseph M. Hilbe
Arizona State University

REFERENCES

Hardin, J. W., & Hilbe, J. M. (2007). *Generalized linear models and extensions*. College Station, TX: Stata Press.

Hilbe, J. M. (2005a). A review of SPSS, part 3: Version 13.0. *American Statistician, 59*, 185–186.

Hilbe, J. M. (2005b). A review of Stata 9.0. *American Statistician, 59*, 335–348.

Hilbe, J. M. (2007a). Statistica 7: An overview. *American Statistician, 61*, 91–94.

Hilbe, J. M. (2007b), Review of ePrint 5 Professional. *American Statistician, 61*, TBD.

CHAPTER 8

CONCLUSION: ROUNDTABLE DISCUSSION

On the last day of the conference, a round table was held with about three dozen participants. A few common themes have arisen and were discussed enthusiastically, One of them was communication between statisticians, methodologists in social sciences, and applied researchers in substantive fields. It was noted that the approach that mathematical statisticians sometimes take is that of patronizing, which is hardly productive, especially when the quantitative fields have developed their own specialized methods. There is also an issue of interests and incentives: statistical departments can make themselves irrelevant to social science researchers if the former concentrate on mathematics while the latter need methods to work with their data. There is a mismatch between the tools used traditionally in statistics and quantitative social sciences, and the push to develop new techniques valued for tenure and promotion. This mismatch is especially troublesome as the temporal lags of penetration, even into relatively close areas, are usually on the order of decades, and the literacy of an average applied researcher is often confined to a two-semester methodological sequence at a social science department.

With the lack of dedicated social science statistics programs and departments, joint appointments in statistics and social sciences seemed to be one of the ways to resolve the problem. Mark Handcock (University of Washington) shared his experiences working at the Center for Statistics and Social Sciences (CSSS) at the University of Washington. The role of CSSS is to provide a middle ground where social science researchers can bring their problems and statisticians can help develop new statistical methods. At the time, CSSS had seven core faculty members who shared time between the Department of Statistics and a social science department. With each faculty, it is clearly delineated which department is considered to be the primary one. It is the department that evaluates the faculty member's research productivity, publications, and grantsmanship and provides the main venue for tenure and promotion. The other department provides accompanying letters of support. John Eltinge (Bureau of Labor Statistics) provided another example of a possible arrangement. At Iowa State University Department of Statistics, another top 10 department in the United States, mathematical statisticians whose main contributions are (or expected to be) theoretical, and aimed for the top statistical journals, have 100% time appointments in statistics, while applied statisticians have joint appointments with the departments of their respective application. Generally, however, joint appointments are difficult to set up and involve extensive discussions with department chairs and deans regarding what is to be valued in the research output of the faculty taking those positions.

Another setting in which statisticians and social sciences often meet in an academic environment is that of statistical consulting. Most large departments have consulting centers and/or offer graduate courses on consulting where initial interactions often occur. A challenge that statisticians frequently face in this situation is that of providing the question of an appropriate level of sophistication. It has been argued at the round table that when there is a prospective for real research, those consulting projects may be converted to full-scale collaborations.

An important issue that a lot of participants connected to was that of funding. Mark Handcock described the funding process at CSSS. The positions of the core faculty of CSSS are funded as hard lines. Additional funding is obtained through federal, state, and campus sources. A part of the overhead money generated from those grants is laid out for seed grants, where statisticians collaborate with social scientists on setting up an agenda at early stages of research. The typical format of those grants is a two-page paper coauthored jointly by a statistician, a social scientist, and a graduate student affiliated with CSSS. This joint work allows the contributors to recognize each others' strengths, and even if the proposal is not funded, this joint work lays the groundwork for getting a clear picture of the substantive problems in a social science field, and of methods that might be available or that need to be developed to solve them.

Peter Bentler (UCLA) commented that similar initiatives at UCLA were less successful, and attributed that fact to the lack of both physical location and the real budget for a stand-alone institution. In turn, the lack of funding might arise from the general treatment of social sciences statistics problems by mathematical statisticians as being less fundable than medical and biological problems that are currently at the forefront of federal funding initiatives. This, in turn, bears on research models and coauthorship patterns: while most papers with applications in medicine and biology carry a long list of authors with probably one or two consulting statisticians, papers in the social sciences tend to be single-authored, and thus methodology is generally limited to whatever tools were available to that sole author.

The structure of positions at Iowa State University became feasible due to the revenue streams from agricultural experiments that started there in the 1940s. Since then, the focus in the design of experiments has shifted to clinical trials. Substantial demand for statistical expertise was generated by the Federal Drug Administration drug approval process requirements whose foundations were laid out in 1940s. Unfortunately, it is not at all clear whether similar streams of funding exist in the social sciences. Some of them might be associated with survey research, which is one of the largest applied statistical areas, as evidenced by the number of members in the Survey Research Methods Section of the American Statistical Association, the largest of all sections. Other streams can be found on interfaces with the biological sciences, such as functional MRI, which enjoys growing popularity in psychology.

The next theme that a lot of participants also deemed crucial was that of education and training: Where do social science methodologists come from now, and where will they be coming from in the future? Obviously, to provide examples and motivation as well as model causal relations, one needs to know the substantive field. But without sufficient statistical background, development of innovative methods will be also hindered. There are very few programs that explicitly offer degrees in social science statistics; the CSSS at the University of Washington in the United States and the social statistics program at Southampton in the UK are the only two examples among top schools. Notably, the number of top programs in quantitative psychology or sociology is quite limited. Again, Mark Handcock reported on the perspective at the CSSS. There, about 20 courses are cross-listed between the statistics department and a social science field, with additional CSSS course number attached. The CSSS also offers camps in mathematics, statistics, and computing for the incoming graduate students before the fall semester. The courses and camps are advertised throughout affiliated departments.

When discussing the foregoing issues, the round-table participants agreed that much of the problem comes from a lack of incentives for interdisplinary collaboration. Different disciplines tend to give different weights to research, training, publication, and grant activities toward tenure and promotion. Publications in journals outside the home department discipline generally count very low. The appearance of strong collaborative centers appears to be related primarily to the good will of university administrators, as it is only the top managers who can provide better incentive structure for collaboration.

The next big theme raised by the round-table participants was the diversity of branches of quantitative social sciences that complement their common underlying principles. The main presentations of the conference covered such areas as structural equation modeling, multilevel modeling, cluster analysis, psychological and educational measurement, social networks, Bayesian methods, missing data, survey statistics, and computing. Contributed and topic-contributed sessions had also touched upon program evaluation and spatial methods, among others. Even though extensive, this list does not cover all of the fields that are of interest and relevance to researchers in the social sciences. Other areas that could have been mentioned are metaanalysis, generalized linear models, latent class analysis and mixture modeling, multidimensional scaling, causal modeling, and multiple-hypothesis testing.

Some common themes can be identified across all of those diverse areas: most important, the issue of identification of latent structures. With rare exceptions, social science studies and the data they generate are observational. The interest of the researcher often lies in unobservable characteristics and their role in the observed phenomena. Thus, structural equation models and item-response theory operate with continuous latent variables; cluster and latent class analysis work with discrete latent variables; and multidimensional scaling and social networks work with hidden geometry of the mental or social space. Program evaluation and treatment effect estimation procedures need to provide counterfactuals in situations where only one out of several possible outcomes can be observed; the estimation problem in question is that of differences between those outcomes. In some situations, identification problems can be resolved by fixing some model parameters at reasonable values, such as having a mean of zero and a variance of 1. In other cases, identification needs to be established from complicated nonlinear procedures.

Another common issue that most quantitative social sciences face is that of software reliance. Applied researchers tend to use a general-purpose software such as SAS or SPSS for basic analysis and one of a handful of highly specialized software packages that implement a specific data analysis procedure. Eventually, researchers find themselves squeezing the data into the models and routines available in these packages, instead of adapting open-ended approaches that lead to the development of new methods. This situation can probably be traced back to the basic mathematical and computer literacy of students coming out of social science graduate programs, and eventually, researchers applying the commonly used methods. Whereas statisticians and econometricians are usually expected to be familiar with one or more statistical packages and proficient enough to program their own routines, this is not necessarily the case for graduate programs in quantitative social sciences.

The round table provided a thoughtful closure to a stimulating conference.

Stanislav Kolenikov
Douglas Steinley
Lori Thombs

INDEX

Statistics in the Social Sciences. By S. Kolenikov, D. Steinley, L. Thombs
Copyright © 2010 John Wiley & Sons, Inc.

Printed in the United States
By Bookmasters